HOW TO PASS

In full COLOUR

INTERMEDIATE 2
BIOLOGY

Harry Hoey

HODDER GIBSON
AN HACHETTE UK COMPANY

Acknowledgements

The Publishers would like to thank the following for permission to reproduce copyright material:

Photo credits
Page 13 (top) Ingram, (bottom) Photolibrary Group; Page 14 Gusto/Science Photo Library; Page 15 © Bettmann/Corbis; Page 64 © Kevin; Fleming/Corbis; Page 65 © Volkmar Brockhaus/zefa/Corbis; Page 91 Juniors Bildarchiv/Alamy; Page 99 (left) Ingram, (centre) www.PurestockX.com, (right) www.PurestockX.com.

Acknowledgements
Extracts from Question Papers are reprinted by permission of the Scottish Qualifications Authority.

Every effort has been made to trace all copyright holders, but if any have been inadvertently overlooked the Publishers will be pleased to make the necessary arrangements at the first opportunity.

Although every effort has been made to ensure that website addresses are correct at time of going to press, Hodder Gibson cannot be held responsible for the content of any website mentioned in this book. It is sometimes possible to find a relocated web page by typing in the address of the home page for a website in the URL window of your browser.

Hachette's policy is to use papers that are natural, renewable and recyclable products and made from wood grown in sustainable forests. The logging and manufacturing processes are expected to conform to the environmental regulations of the country of origin.

Orders: please contact Bookpoint Ltd, 130 Milton Park, Abingdon, Oxon OX14 4SB. Telephone: (44) 01235 827720. Fax: (44) 01235 400454. Lines are open 9.00–5.00, Monday to Saturday, with a 24-hour message answering service. Visit our website at www.hoddereducation.co.uk. Hodder Gibson can be contacted direct on: Tel: 0141 848 1609; Fax: 0141 889 6315; email: hoddergibson@hodder.co.uk

First published in 2006 by
Hodder Gibson, an imprint of Hodder Education,
An Hachette UK Company,
2a Christie Street
Paisley PA1 1NB

This colour edition first published 2009

Impression number 5 4 3
Year 2012 2011

Cover photo © Royal Geographical Society/Alamy
Cartoons © Moira Munro 2005, 2008
Typeset in 10.5 on 14pt Frutiger Light by Phoenix Photosetting, Chatham, Kent
Printed in Dubai

A catalogue record for this title is available from the British Library

ISBN-13: 978 0340 974 100

CONTENTS

INTRODUCTION

This book has been produced to form a study guide to help you pass Intermediate 2 Biology. To pass Intermediate 2 Biology at a high grade you have to develop many different skills. This book has been written to meet all the requirements of the course work. These include the learning outcomes of the course (course content), skills in Problem Solving and Practical Abilities and skills in answering examination questions.

In each chapter, the learning outcomes of the course content have been clearly written. Work tasks are used to help reinforce your understanding of the learning outcomes contained within the course content. It would be very helpful if you had a copy of the Arrangements Document for Intermediate 2 Biology, as it contains the course content that is tested in the National Examinations together with other details of the course. Your school may already have supplied this; however, a copy, can be downloaded or accessed on the SQA website: www.sqa.org.uk.

Hints and Tips

One of the tips given in each chapter is that you should produce 'flash cards' for the words and phrases that are highlighted in **bold** throughout the text. 'Flash cards' are an excellent learning aid and can be used by an individual, by pairs or by a group.

Written on one side of the card is a named biological process or structure and on the other is a definition of the process or a description of the function.

The two sides of a typical 'flash card' are shown in the diagram below:

Side 1	Side 2
Cerebellum	**Centre for balance and co-ordination of movement**

Both sides are used to ask a question.

Side 1. What is the function of the cerebellum?
Side 2. Name the structure that is the 'Centre for balance and co-ordination of movement'.

Cards can be cut out from A4 paper to a rectangular size of approximately 10.5 × 3.0 cm. To keep the cards for a chapter together, punch a hole at the top left-hand corner and keep them together in a key ring. Carry the key ring with you and you can study the key words for that chapter at any time and anywhere.

***Hints** and **Tips** continued* ➢

Hints and Tips continued

Sometimes you will have to interpret the details of the card. If a word is in bold in the text such as 'produce', look at the heading for the area. If the heading is 'carbon fixation', then on one side you would write 'What is produced in carbon fixation' and on the other side would be 'glucose'.

At the end of the course content for each chapter there is an end of chapter test. Questions are set to the standard of the external examination. At the end of each chapter, a clear commentary on how to answer the end of chapter questions is given.

To help develop skills in answering these questions, there are chapters on skills in answering multiple choice questions, structured questions, problem solving questions, practical abilities questions and extended writing questions. This will help you in answering all the different question types that appear in the National Examination.

On pages 3–7 there are revision grids for each of the Units. Use the grids in your revision for:

1 End of Unit Tests (NABs), and
2 the National Examination.

Overall, this book, if used correctly, will help you to become aware of what is required in order to gain a pass at a high grade in the National Examination. I hope that you enjoy using the book and that it helps you to be better prepared for your examination.

Harry Hoey

How good am I? Scale 1–10	Revision Topics for Unit 1	Revision check	
	Unit 1 Living cells	1	2
	Cell structure and function		
	(i) Similarities and differences between animal, plant and microbial cells		
	(ii) Function of cell structures		
	(iii) Commercial & industrial uses of cells		
	Diffusion and osmosis in plant and animal cells		
	(i) (ii) Diffusion and its importance to cells		
	(iii) (iv) Osmosis and its effects in cells		
	Enzyme action		
	(i) Properties of catalysts and enzymes		
	(ii) Specificity of enzymes for their substrate		
	(iii) Degradation and synthesis		
	(iv) Factors affecting enzyme activity		
	Aerobic and anaerobic respiration		
	(i) Glucose as a source of energy in the cell		
	(ii) Role of ATP		
	(iii) Comparison of energy yield and products of aerobic and anaerobic pathways		
	Photosynthesis		
	1. Energy fixation		
	(i) Sunlight as the source of energy		
	(ii) Equation for photosynthesis		
	(iii) Photolysis and carbon fixation		
	(iv) Conversion of glucose to other carbohydrates		
	2. Factors affecting rate of photosynthesis		
	(i) Limiting factors		
	(ii) Production of early crops in horticulture		

How good am I? Scale 1–10	Revision Topics for Unit 2	Revision check	
	Unit 2 Environmental Biology and Genetics	1	2
	Ecosystems		
	1. Energy flow		
	(i) Components of an ecosystem		
	(ii) Food chains and food webs		
	(iii) Flow of energy through food chains in an ecosystem		
	2. Factors affecting the variety of species in an ecosystem		
	(i) Importance of biodiversity at species level		
	(ii) Factors affecting biodiversity		
	(iii) Behavioural adaptations in animals and their adaptive significance		
	Factors affecting variation in a species		
	1. Fertilisation		
	(i) Continuous and discontinuous variation		
	(ii) Gamete (sex cell) production		
	(iii) Fusion of nuclei		
	2. Genetics		
	(i) Division of nucleus in gamete production (meiosis)		
	(ii) Importance of chromosome structure to an organism's characteristics Relationship between DNA sequence and protein synthesised Relationship between cell proteins and organism's characteristics		
	(iii) Chromosome numbers in different species		
	(iv) Sex determination		
	(v) Characteristics controlled by alleles		
	(vi) Terms used in genetics to the F_2 generation		
	(vii) Monohybrid crosses		
	(viii) Proportions and ratios of phenotypes of the F_1 and F_2 offspring		

How good am I? Scale 1–10	Revision Topics for Unit 2	Revision check	
	(ix) Co-dominance		
	(x) Polygenic inheritance		
	(xi) Environmental impact on phenotype		
	3. Selection		
	(i) Natural selection		
	(ii) Selective breeding		
	(iii) Genetic engineering		

How good am I? Scale 1–10	Revision Topics for Unit 3	Revision check	
	Unit 3 Animal Physiology	1	2
	Mammalian nutrition		
	1. Breakdown of food		
	(i) Food groups		
	(ii) Food tests and energy content of food		
	(iii) The need for digestion		
	2. The structure and function of the alimentary canal		
	(i) Mouth, salivary glands and oesophagus		
	(ii) The role of the stomach		
	(iii) The role of the small intestine, pancreas, liver and gall bladder		
	(iv) Absorption of food through the wall of the small intestine		
	(v) The fate of absorbed material		
	(vi) The role of the large intestine, rectum and anus in digestion		
	Control of the internal environment		
	(i) Structure of the human urinary system		
	(ii) The role of the mammalian kidney		
	(iii) Negative feedback control by ADH		
	(iv) Osmoregulation in marine and freshwater fish		
	Circulation and gas exchange		
	1. Structure and function of heart and blood vessels		
	(i) The structure of the heart		
	(ii) Blood vessels		
	(iii) Circulation of blood		
	2. Composition and functions of blood		
	(i) Function of RBCs and plasma in transport		
	(ii) Function of haemoglobin		

How good am I? Scale 1–10	Revision Topics for Unit 3	Revision check	
	(iii) Functions of macrophages and lymphocytes		
	3. Structure and function of lungs in gas exchange and the capillary network		
	(i) Internal structure of lungs and features which make them efficient gas exchange structures		
	(ii) Features of capillary network which allow efficient gas exchange in tissues		
	Sensory mechanisms and processing information		
	1. The structure and function of the brain		
	(i) Function of cerebrum, cerebellum, medulla and hypothalamus		
	(ii) Sensory and motor areas of cerebrum		
	2. Structure and function of the nervous system		
	(i) Brain, spinal cord and nerves and role of Central Nervous System (CNS)		
	(ii) Reflex action and the reflex arc		
	(iii) Temperature regulation as a negative feedback mechanism		

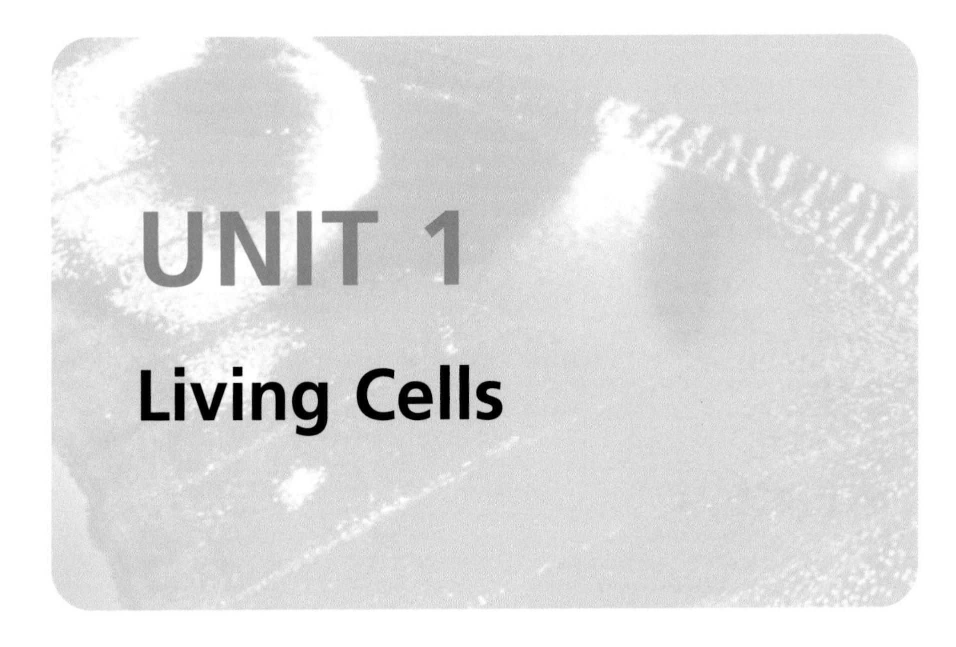

UNIT 1
Living Cells

Chapter 1

CELL STRUCTURE AND FUNCTION

Similarities and Differences between Animal, Plant and Microbial Cells

Figure 1.1 show structures present in a human cheek epithelial cell (animal), a leaf mesophyll cell (plant) and a yeast cell (microbial).

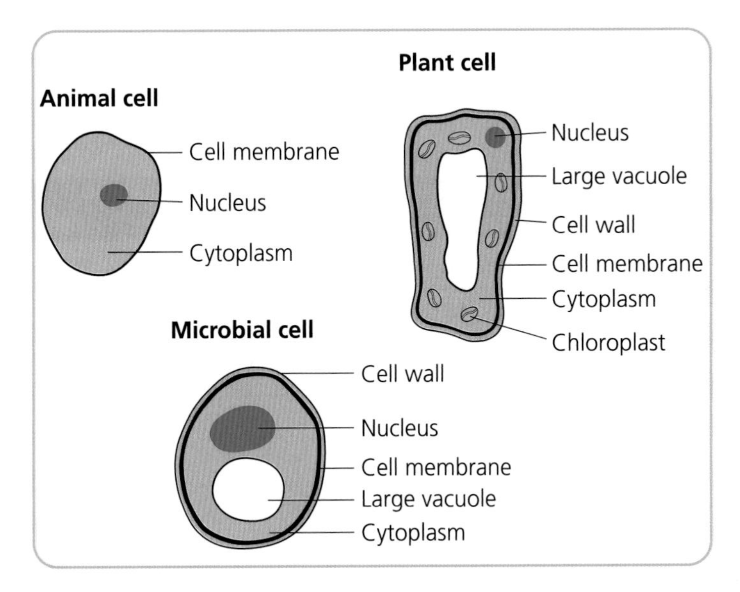

Figure 1.1

Key Points

- ***Cheek cell versus mesophyll cell***

  Similarities: Presence of a cell membrane, cytoplasm and nucleus.

  Differences: Absence of a cell wall, large vacuole and chloroplasts from a cheek cell.

- ***Cheek cell versus yeast cell***

  Similarities: Presence of a cell membrane, cytoplasm and nucleus.

  Differences: Absence of a cell wall and large vacuole from a cheek cell.

- ***Mesophyll cell versus yeast cell***

  Similarities: Presence of a cell membrane, cytoplasm, nucleus, cell wall and large vacuole.

  Difference: Absence of chloroplasts from a yeast cell.

Hints and Tips

If a cell wall is present it cannot be an animal cell.

If chloroplasts are present it cannot be an animal cell or a yeast cell.

If large vacuoles are present it cannot be an animal cell (NB animal cells can have small vacuoles).

You may be asked to calculate the length or diameter of cells. Use the fact that 1000 micrometers = 1 millimetre. To convert millimetres to micrometers multiply by 1000 and to convert micrometers to millimetres divide by 1000.

Stop Think Learn

1 Copy and complete this table to show structures that are present or absent from each cell type (the nucleus is shown for you).

Structure	Cheek epithelial cell		Leaf mesophyll cell		Yeast cell	
	Present	Absent	Present	Absent	Present	Absent
Nucleus	Present		Present		Present	

2 Without looking at Figure 1.1, draw and label a cheek epithelial cell, a leaf mesophyll cell and a yeast cell to show their structure.

Function of Cell Structures

Key Points

- **Cell membrane:** Controls the movement of materials into and out of the cell.
- **Nucleus:** Controls all the activities of the cell to include cell metabolism and cell division (**contains the genetic material**).
- **Cytoplasm:** This is the area in which the activities of the cell take place.
- **Cell wall:** Gives the cell shape and prevents the cell from bursting when water is entering by osmosis.
- **Large vacuole:** Used for storage of cell materials.
- **Chloroplast:** Absorb light energy used to produce glucose in photosynthesis.

Stop Think Learn

Using suitable headings, construct a table to show the function of all the structures present in a leaf mesophyll cell (check your answers against the notes).

Commercial and Industrial uses of Cells

Key Points

- The commercial and industrial uses of cells depend on cell respiration.
- Respiration takes place within living cells. Respiration is the release of energy from the breakdown of glucose and is of two main types:
  1. Aerobic respiration requires the presence of oxygen.
  2. Anaerobic respiration requires that oxygen is absent. (Remember, 'anaerobic' contains the letter 'n', so **no** oxygen is present.)
- In anaerobic respiration in yeast, glucose is broken down to form the end products ethanol and carbon dioxide (anaerobic respiration is also referred to as fermentation).

Word equation

glucose ⟶ carbon dioxide + ethanol (alcohol)

Examples

Bread making

Carbon dioxide gas released in anaerobic respiration of yeast is used to make the dough rise.

Figure 1.2

Alcohol production

In beer (brewing) and wine production, the ethanol formed in anaerobic respiration of yeast is the source of alcohol.

Figure 1.3

Examples *continued* ➢

Examples continued

Antibiotic production

Many different fungal species are used to produce a wide range of antibiotics.

Many bacteria can cause disease.

Antibiotics destroy bacteria.

Having a wide range of antibiotics means that many different types of bacteria can be destroyed.

Some species of bacteria, particularly in hospitals, are no longer destroyed by some of the antibiotics. These are resistant strains of bacteria.

The number of resistant strains of bacteria is on the increase and this is as a result of the overuse of antibiotics.

Yoghurt production

Yoghurt is made from milk. Specialised bacteria, in a fermentation process, convert lactose sugar in milk to lactic acid. The increase in lactic acid causes the milk proteins to coagulate.

Word equation

lactose ⟶ lactic acid

Figure 1.4

Alternative fuel production

Fuels are used as the energy source for all of man's activities.

Natural fuels include coal, natural gas and oil, together with all their products such as petrol, diesel, etc.

Natural fuels are finite and as they are used up they are not renewed.

Alternative fuels are produced by the activities of living cells and these are renewable.

Biogas (methane gas) production

Waste products (household refuse, paper, etc.) are used as a food source for bacteria. The anaerobic respiration (fermentation) of the bacteria breaks the waste down to methane gas. The gas produced is bottled.

Examples *continued* ➢

Examples continued

Gasohol

In anaerobic respiration (fermentation) by yeast, sugars are broken down and ethanol (alcohol) is formed. The alcohol is mixed with petrol to produce gasohol.

(The engines of some cars are now designed to use pure alcohol.)

Figure 1.5

Hints and Tips

Make 'flash cards' for the words and phrases in **bold**.

Stop Think Learn

Copy and complete this table to show details of commercial and industrial uses of cells. Bread making is shown for you. (Check your answers against the notes.)

Commercial and industrial uses of cells	Details
Bread making	Carbon dioxide gas released in anaerobic respiration of yeast is used to make the dough rise.
Alcohol production	

Hints and Tips

For examination-style questions, the number beside the question shows the mark allocation.

All multiple choice questions are allocated one mark each.

Before answering multiple choice questions revise Chapter 14.

Before answering structured questions revise Chapter 15.

Before answering extended writing questions revise Chapter 16.

Check your answers from the commentary given at the end of each chapter.

Questions

Examination style questions

Questions 1 and 2 are based on Figure 1.6.

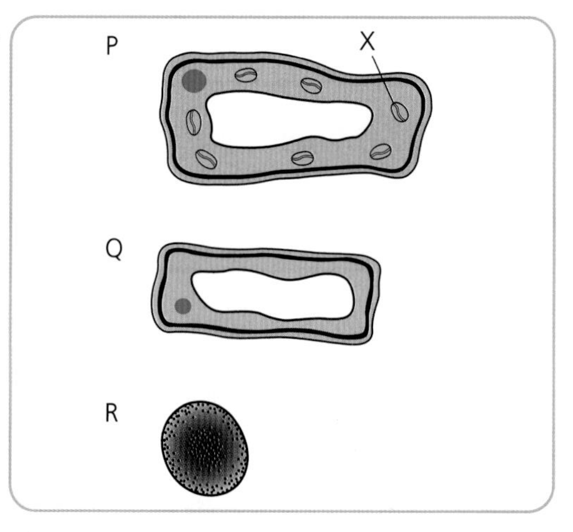

Figure 1.6

1 The plant cells are:

A P only.

B P and Q only.

C P and R only.

D R only.

2 The function of structure X is to:

A control all activities of the cell.

B store starch.

C produce glucose using light energy.

D release energy from glucose.

3 Figure 1.7 shows plant cells as seen under a microscope at a magnification of ×100

The diameter of the field of view is 200 micrometers. The average width of each cell in micrometers is:

A 0.05. B 0.5. C 5.0. D 50.0.

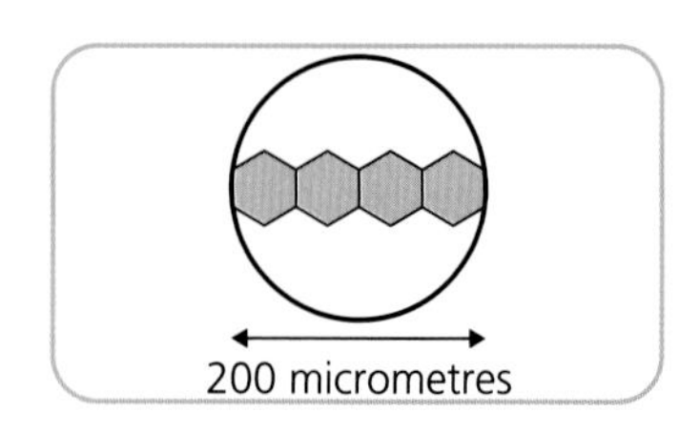

Figure 1.7

Questions continued

4 Figure 1.8 shows sections through a cheek epithelial cell, a leaf mesophyll cell and a yeast cell.

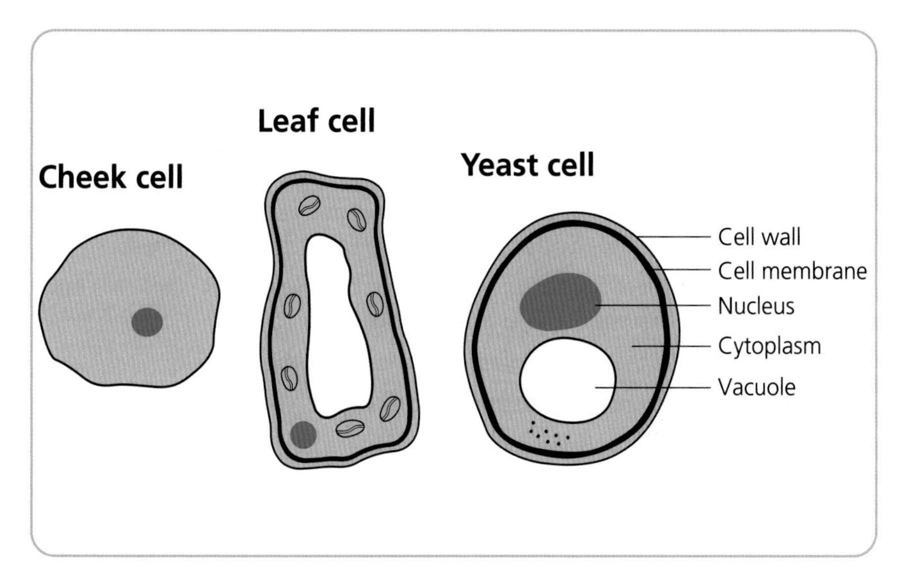

Figure 1.8

a) Identify three structures present in the yeast cell which are also present in the cheek cell. *(1)*

b) Identify two structures present in both the yeast cell and mesophyll cell which are absent from the cheek cell. *(1)*

c) Identify a structure present in the leaf cell which is absent from the other two cells. *(1)*

d) State the function of the following structures:

(i) cell membrane, (ii) vacuole, (iii) nucleus. *(2)*

5 Yeast is a micro-organism used in the production of bread. Name one other type of micro-organism and an associated product. *(1)*

6 Describe five examples of the commercial and industrial uses of cells. *(5)*

Answers

Answers to examination style questions with commentary

1 You should know that plant cells have a cell wall. **Answer = B**

2 Structure X is a chloroplast. You should know that this is the site for photosynthesis in which light energy is used and glucose is made. **Answer = C**

3 Diameter magnified by ×100 is 200 micrometres. Therefore, true diameter is 200 divided by 100 = 2 micrometres. Count the number of cells within the diameter of the field = 4 cells. Therefore, average width of cell = 2 divided by 4 = 0.5 micrometres. **Answer = B**

4 a) This is asking you to identify structures present in both yeast and animal cells. **Answer = cell membrane, nucleus, cytoplasm.** All 3 correct = 1 mark

b) This is asking you to identify structures present in both yeast and plant cells but not in animal cells. **Answer = cell wall, vacuole.** Both correct = 1 mark

c) You have to recognise that **chloroplasts** are present in the leaf cell.

d) You should know the functions of structures:

(i) cell membrane – controls the movement of materials into and out of the cell.

(ii) vacuole – used for storage of cell materials.

(iii) nucleus – controls all the activities of the cell including cell metabolism and cell division.

3 correct = 2 marks. 2 correct = 1 mark.

5 Micro-organisms other than yeast would be other fungi or bacteria. Fungi used in antibiotic production. Bacteria used in yoghurt or biogas (methane) production. **Answer = any one from the three above.**

6 Describe **five** examples of the commercial and industrial uses of cells. In extended writing questions you **must** identify all the parts that make up the question. These are:

Five examples of the commercial and industrial uses of cells must be given. **Each** example must be described clearly.

Marking instructions

A1 Anaerobic respiration/fermentation in yeast produces carbon dioxide which causes dough to rise.

A2 Alcohol produced in anaerobic respiration/fermentation by yeast is used in the brewing/wine making industries.

A3 Fungi used to produce a wide variety of antibiotics.

A4 Bacteria convert lactose to lactic acid in yoghurt production.

A5 Bacteria produce biogas/methane from waste products.

A6 Alcohol produced by yeast is mixed with petrol to produce gasohol.

1 mark for each. Maximum mark = 5.

Chapter 2

DIFFUSION AND OSMOSIS IN ANIMAL AND PLANT CELLS

Diffusion and Importance of Diffusion to Cells

Diffusion

Key Points

- A **solution** is made up of a solute and a solvent.
- In a glucose solution, glucose is the solute and water is the solvent.
- The solute concentration is the weight of solute dissolved in a given volume of solvent.

If 10 g of glucose is dissolved in 1 litre of water the concentration of the solute is 10g/l.

- **Diffusion** takes place if there is a difference in solute concentration between two areas.
- One area must have a **higher solute** concentration than the other.
- In a cell these areas will be the fluid surrounding the cell and the contents of the cell.
- In cells, the two areas will be separated by the cell membrane (see Figure 2.1).
- Figure 2.2 shows that the area outside the cell has a higher solute concentration than the area inside the cell.
- The difference in concentration is described as a concentration gradient (direction of gradient shown by an arrow).
- The dissolved solute moves down the concentration gradient from the area of higher solute concentration to the area of lower solute concentration (shown by an arrow).

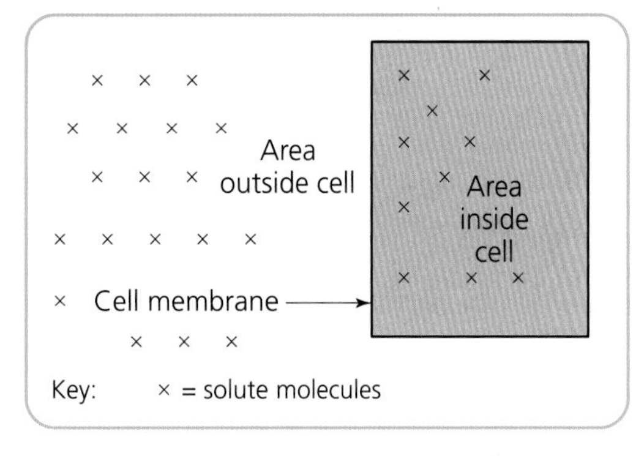

Figure 2.1 The process of diffusion

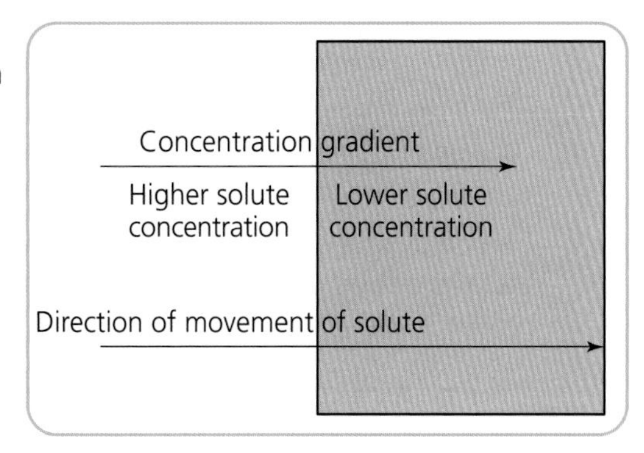

Figure 2.2 The process of diffusion in terms of a concentration gradient

Definition of diffusion

Diffusion is the movement of substances from an area of higher solute concentration to an area of lower solute concentration down a concentration gradient.

Examples of substances that enter or leave cells by diffusion

The arrows in Figure 2.3 show the path of diffusion down the concentration gradient from the area of higher solute concentration to the area of lower concentration in an animal cell and a leaf mesophyll cell in the light.

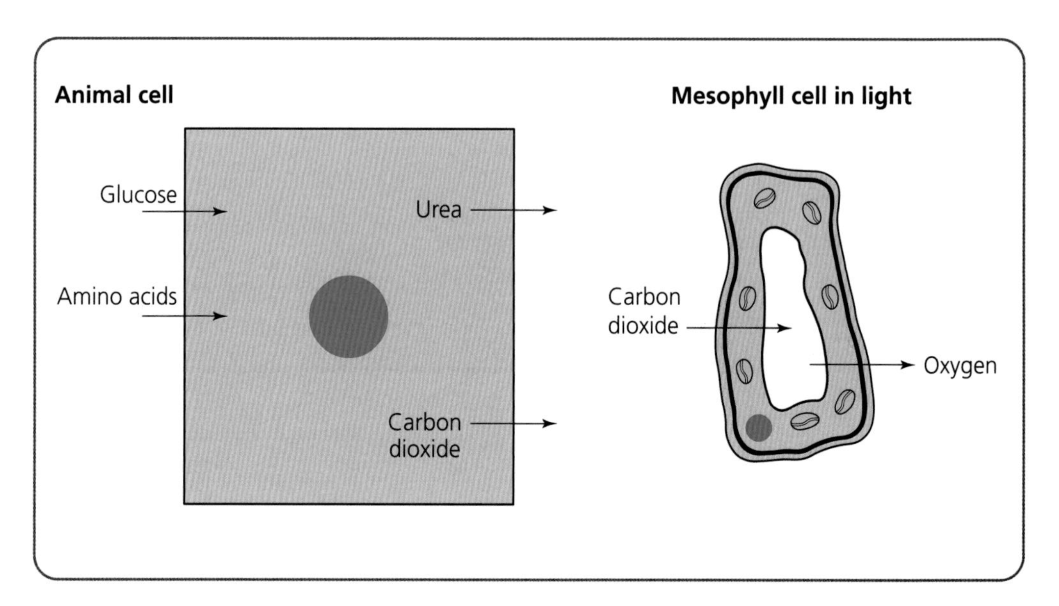

Figure 2.3

Importance of diffusion to cells

Key Points

- Glucose and oxygen that enter cells by diffusion are needed for respiration.
- Respiration is important because the energy released is used for all the activities of the cell.
- Urea and carbon dioxide leave cells by diffusion. This is important because high concentrations of urea and carbon dioxide are poisonous to the cells.
- Carbon dioxide enters leaf mesophyll cells by diffusion in the light. This is important because carbon dioxide is a raw material for photosynthesis and is required for the production of glucose.

To make certain that the concentration gradient and thus the direction of movement for diffusion is correct use the **solute concentration gradient triangle** (Figure 2.4).

Example

A cell is surrounded by a solution with a glucose concentration of 5 g/l. The cytoplasm of the cell is at a glucose concentration of 2 g/l.

State the direction of diffusion of glucose and explain your answer in terms of a concentration gradient.

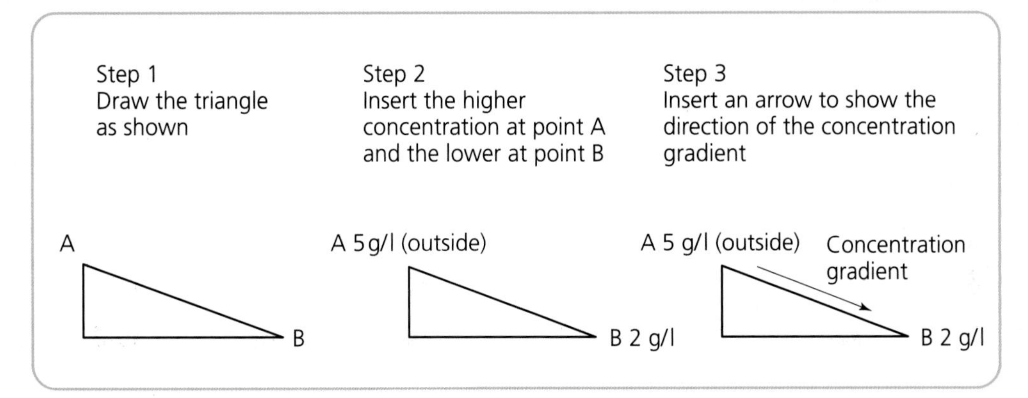

Figure 2.4

Answer

Glucose will diffuse into the cell. It is because there is a higher glucose concentration outside the cell and glucose will move down the concentration gradient.

Stop Think Learn

1 How would a glucose solution with a concentration of 5 g/l be prepared?

2 Use the concentration gradient triangle to find the direction in which the following solutes will diffuse.

	Concentration outside the cell	Concentration inside the cell
a)	Glucose of 10 g/l	Glucose of 5 g/l
b)	Oxygen of 0.1 mm^3/l	Oxygen of 0.1 mm^3/l
c)	Urea of 0.01 g/l	Urea of 0.1 g/l

3 Copy and complete the table below with an additional three rows to show the pathway of diffusion of glucose, urea, carbon dioxide and amino acids and the importance of these to the cell. Glucose has been completed for you. (Check your answers against the notes.)

Substance diffusing	Pathway	Importance to the cell
Glucose	Into cell	Glucose is needed for respiration and the energy released is used for all the activities of the cell.

Osmosis and Osmotic Effects in Animal and Plant Cells

Osmosis

Key Points

- The cell membrane controls the movement of materials into and out of the cell. Such a membrane is described as being a **selectively permeable membrane (SPM)**.
- The SPM has pores (see Figure 2.5.)

Figure 2.5 shows the relative size of sucrose and water molecules.

- The sucrose molecules are too large to pass through the pores in the membrane whereas the water molecules are small enough to pass through.

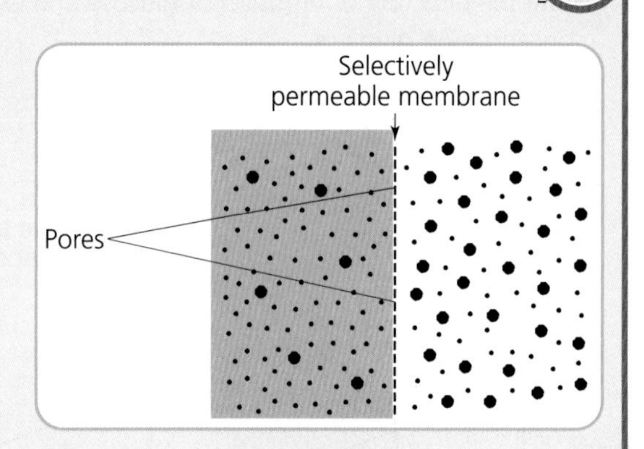

Figure 2.5
Key ● = sucrose molecules • = water molecules

- Water molecules can pass through the SPM but sucrose molecules cannot.
- Osmosis is a 'special case' of diffusion. Osmosis refers to the diffusion of water molecules. The rules for diffusion of water are the same as for the diffusion of solutes.
- There has to be a difference in water concentration between two areas.
- One area must have a higher water concentration than the other.
- In cells, the two areas will be separated by the cell membrane that acts as a SPM (see Figure 2.6).
- Figure 2.6 shows that outside the cell has a higher water concentration than inside.
- The difference in concentration is described as a **water concentration gradient**. An arrow in Figure 2.6 shows the direction of the gradient.

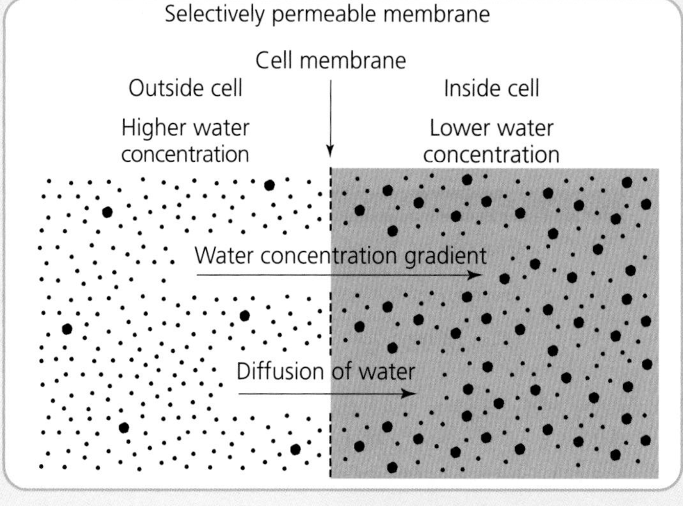

Figure 2.6

Key Points continued ➢

Key Points continued

- Water moves by diffusion down the water concentration gradient from the area of higher water concentration to the area of lower water concentration. An arrow in Figure 2.6 shows the direction of diffusion.

Definitions of osmosis

Osmosis is the diffusion of water from an area of higher water concentration to an area of lower water concentration across a SPM.

OR

Osmosis is the diffusion of water across a SPM as the result of a water concentration gradient.

Hints and Tips

When working out the direction of osmosis always use water concentrations. To do this you will often have to convert solute concentrations to water concentrations.

Example

You are given two solutions: a 10% sugar solution and a 20% sugar solution.

If the solution is 10% sugar then it is 90% water and if it is 20% sugar then it is 80% water. You now have the water concentrations of the two solutions.

You are given two solutions: a 5 g/l sugar solution and a 10 g/l sugar solution.

Five grams of sugar will take up less space within 1 litre than 10 g. Therefore, the 5 g/l must have a higher water concentration than the 10 g/l solution.

To make certain that you get the water concentration gradient and thus the direction of movement of water correct use the **water concentration gradient triangle**.

Example

A cell is surrounded by a solution with a concentration of 5 g/l. The cytoplasm of the cell is at a concentration of 2 g/l.

State the direction of osmosis and explain your answer in terms of a water concentration gradient.

How to answer

1 Translate solute concentration into water concentration.

2 g of solute takes up less space in 1 litre than 5 g. Therefore 2 g/l has the higher water concentration.

2 Construct the water concentration gradient triangle (Figure 2.7).

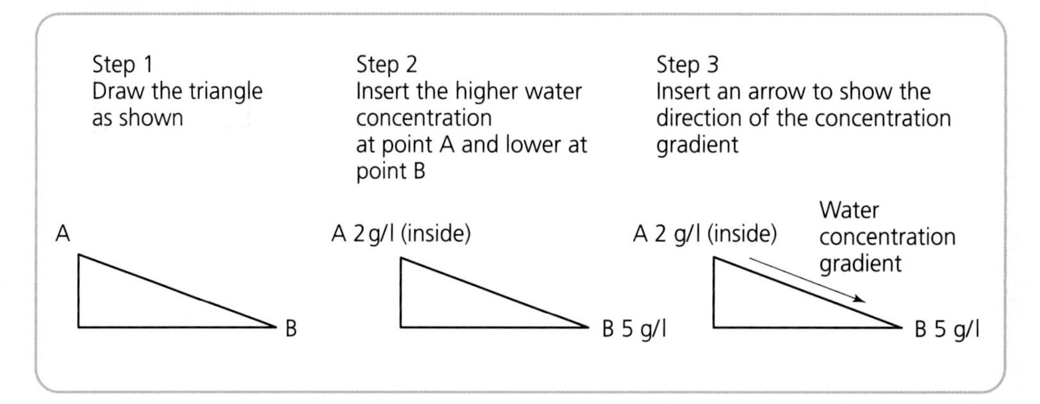

Figure 2.7

Answer

Water will move out of the cell by osmosis. It is because there is a higher water concentration inside the cell and water will move down the water concentration gradient.

Stop Think Learn

Use the water concentration gradient triangle to find out the direction of osmosis in the following situations:

	Concentration outside the cell	Concentration inside the cell
a)	Glucose of 10 g/l	Glucose of 5 g/l
b)	10% glucose solution	20% glucose solution
c)	5% glucose solution	5% glucose solution

Osmotic effects in animal and plant cells

Tonicity is used to compare the solute concentration of two solutions.

A **hypertonic solution** has a higher solute concentration than the solution it is being compared with.

A **hypotonic solution** has a lower solute concentration than the solution it is being compared with.

An **isotonic solution** has the same solute concentration as the solution it is being compared with.

Hints and Tips

Always convert tonicity terms to water concentration and then use the **water concentration gradient triangle.**

Hypertonic solution = higher solute concentration = lower water concentration

Hypotonic solution = lower solute concentration = higher water concentration

Isotonic solution = same solute concentration = same water concentration

Effect of placing animal cells in solutions of different tonicity

Appearance of cells before immersion			
Solution immersed in	Hypertonic	Hypotonic	Isotonic
Water concentration of solution	Lower than cell	Higher than cell	Same as cell
Water concentration triangle	Inside higher water conc. Water conc. gradient Outside lower water conc.	Outside higher water conc. Water conc. gradient Inside lower water conc.	No triangle as there is no water concentration gradient
Direction of osmosis	Water moves out	Water moves in	No osmosis
Effect on appearance of cell	Cell volume decreased; cell shrinks	Cell volume increased; cell swells and bursts	Cell remains unchanged
Appearance of cell after immersion			

Effect of placing plant cells in solutions of different tonicity

Appearance of cells before immersion			
Solution immersed in	Hypertonic	Hypotonic	Isotonic
Water concentration of solution	Lower than cell	Higher than cell	Same as cell
Water concentration triangle	Inside higher water conc. Water conc. gradient Outside lower water conc.	Outside higher water conc. Water conc. gradient Inside lower water conc.	No triangle as there is no water concentration gradient
Direction of osmosis	Water moves out	Water moves in	No osmosis
Effect on appearance of cell	Cell volume decreased; size of vacuole decreased; cell membrane detached from cell wall	Cell volume increased; size of vacuole increased; cell wall pushed out	Cell remained unchanged
Appearance of cell after immersion			
Terms used to describe appearance of cells	Plasmolysed	Fully turgid	Cells are turgid but not fully turgid
Other points	Space between cell wall and cell membrane is filled with the surrounding solution as cell wall is permeable	A cell is turgid when the cell contents push out against the cell wall. The vacuole has increased in size	

Stop Think Learn

Copy and complete the following table to describe the effect on the appearance of the cells after being placed in the different solutions.

Type of cell	Effect on appearance of cell after being placed in solution		
	Hypertonic	*Hypotonic*	*Isotonic*
Animal			
Plant			

Hints and Tips

Make 'flash cards' for the words and phrases in **bold**.

In problems with plant tissue, such as potato tissue, in which there is a loss in weight when placed in a solution, then the tissue has lost water. Water must be leaving the cells of the tissue by osmosis. NB The solution is hypertonic.

In problems with plant tissue, such as potato tissue, in which there is a gain in weight when placed in a solution, then the tissue has gained water. Water is entering the cells of the tissue by osmosis. NB The solution is hypotonic.

Only plant cells can become plasmolysed or turgid. These descriptions refer to cells that have a cell wall.

Animal cells can never become plasmolysed nor turgid as they do not have a cell wall.

Questions

Examination style questions

1 Figure 2.8 shows a unicellular organism *Paramecium* that lives in fresh water.

a) Name the process by which oxygen moves from the water into the organism. *(1)*

b) Name a substance that moves from the organism into the water. *(1)*

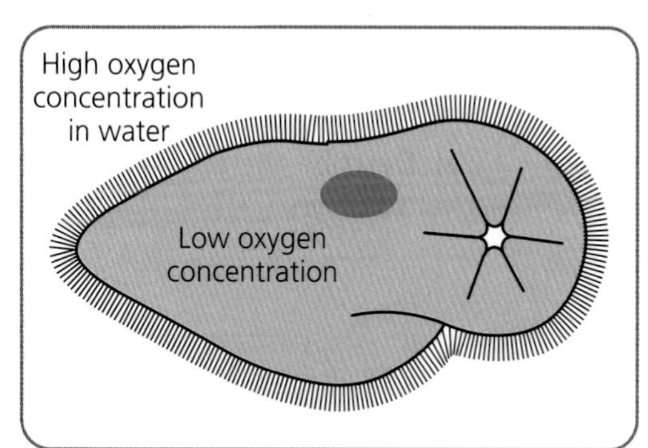

Figure 2.8

Questions continued

2 Figure 2.9 represents a section of human tissue showing an exchange of chemical substances between body cells and blood.

a) Name and describe the process by which carbon dioxide moves out of the body cells into the blood. *(2)*

b) Why is it important that carbon dioxide is removed from the body cells? *(1)*

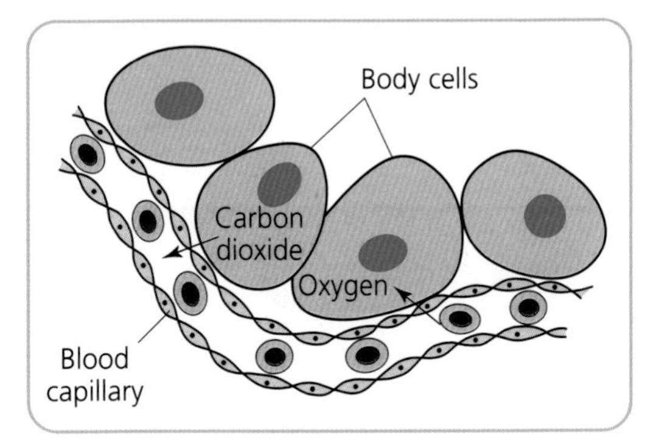

Figure 2.9

3 Three discs were cut from the same potato and were placed in three salt solutions of different concentrations.
After 30 minutes, the discs were removed from the solutions and the cells examined under a light microscope. A cell from each disc is shown in Figure 2.10.

a) Identify the cell which was placed in
(i) a hypertonic solution, (ii) an isotonic solution. *(1)*

b) Name the process which causes the difference in appearance of the cells. *(1)*

c) What name is used to describe the condition of cell C? *(1)*

d) Name the cell structure which prevents plant cells from bursting. *(1)*

e) Describe the appearance of red blood cells when placed in a hypertonic solution. *(1)*

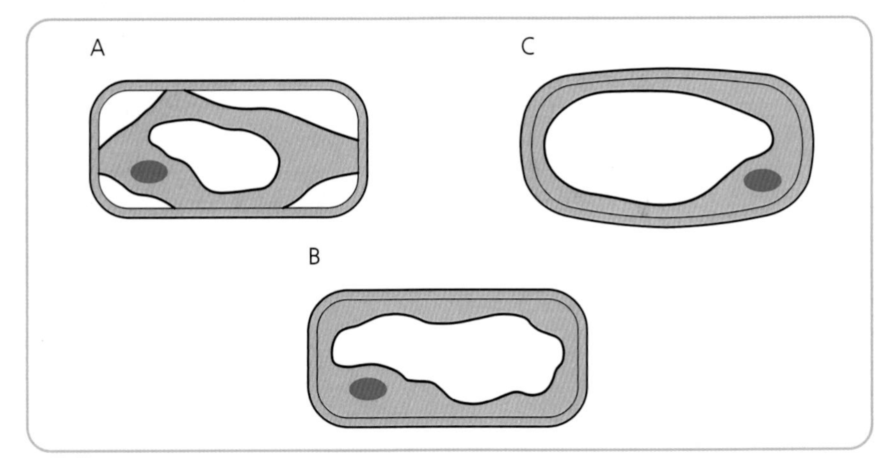

Figure 2.10

4 Figure 2.11 shows the appearance of human red blood cells in an isotonic solution.

Describe and explain the events that would take place if these red blood cells were transferred to pure water. *(5)*

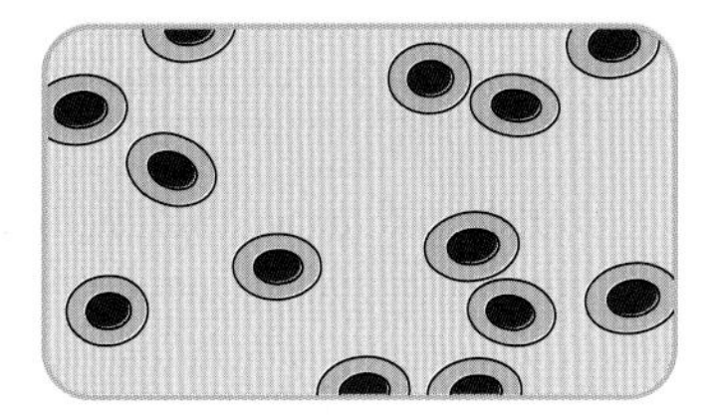

Figure 2.11

Answers

Answers to examination style questions with commentary

1 a) This is movement of oxygen from an area of higher oxygen concentration in the water to an area of lower oxygen concentration inside the organism.

Answer = Diffusion

b) All living cells must respire. The waste product of respiration is carbon dioxide.

Carbon dioxide concentration will become higher inside the organism than outside and will diffuse out from the organism.

Answer = Carbon dioxide

2 a) Draw the triangle (see Figure 2.12).

You should know that gases move from an area of higher concentration to an area of lower concentration in a process called diffusion.

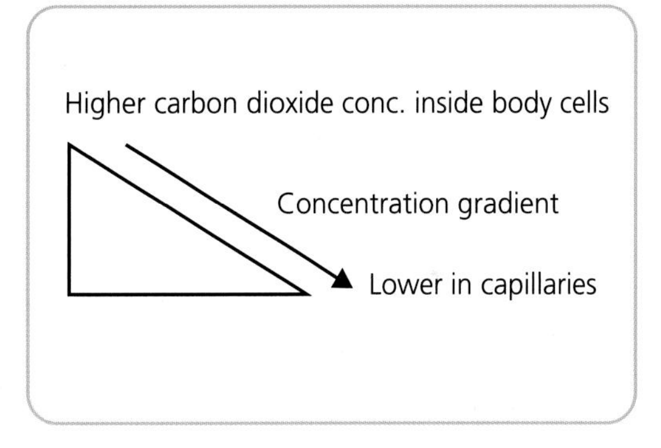

Figure 2.12

Answer = Diffusion

Description

Carbon dioxide moves down a concentration gradient from an area of higher concentration inside the cells to a lower concentration in the capillary.

b) If not removed the concentration of carbon dioxide would increase and could reach toxic levels.

3 a) You should know that a hypertonic solution has a lower water concentration and that water moves down the water concentration gradient. Draw the triangle (see Figure 2.13).

As water moves out of the cell the cell would plasmolyse.

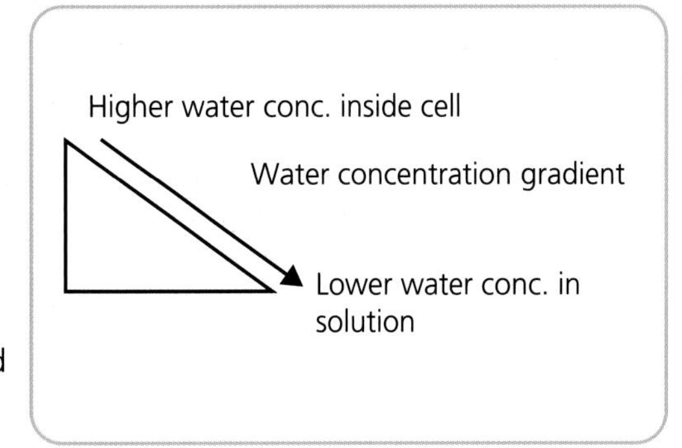

Figure 2.13

Answers continued

You should know that isotonic solutions have the same water concentration and as a result there is no overall movement of water in or out. The cell is unchanged.

Answer = Hypertonic solution = Cell A, Isotonic solution = Cell B. Both for 1 mark

b) You should know that diffusion of water is called osmosis.

Answer = Osmosis

c) In cell C, the vacuole is large and the cell wall is being pushed out.

Answer = Turgid (fully)

d) This is knowledge of the function of the cell wall in plant cells.

Answer = Cell wall

e) You should know that a hypertonic solution has a lower water concentration.

Use the same triangle in Figure 2.13. You know that water will leave the red blood cells and because these are animal cells it cannot become plasmolysed; it simply gets smaller in size.

Answer = The red blood cells shrink in size

4 Describe and explain the events that would take place if these red blood cells were transferred to pure water.

In extended writing questions you **must** identify all the parts that make up the question. These are:

- Describe the events that would take place after transfer.
- Explain these events.

Marking instructions

A1 Water enters the cells.

A2 The cells increase in volume/swell as water enters.

A3 The cell continues to swell and bursts.

Maximum 2 marks

B1 The pure water is hypotonic to the red blood cells OR the pure water has a higher water concentration than the blood cells.

B2 The cell membrane acts as a selectively permeable membrane.

B3 Water moves from the higher water concentration outside the cells to the lower water concentration inside the cells.

B4 The movement of water is called osmosis.

Maximum 3 marks.

Maximum total = 5 marks.

Chapter 3

ENZYME ACTION

Properties of Catalysts and Enzymes

Key Points

- All **catalysts** have the following properties:
  Catalysts lower the energy input that is required to cause a chemical reaction.
  Catalysts speed up the rate of chemical reactions.
  Catalysts remain unchanged after taking part in a chemical reaction.
  Catalysts carry out the same reaction many times.
- **Enzymes** are described as biological catalysts. Other characteristics of enzymes are:
  Living cells synthesise enzymes.
  Enzymes are proteins.
  Enzymes are required for the functions of all living cells.

Stop Think Learn

1. Copy and complete Figure 3.1 by showing a property that is common to all catalysts in each of the boxes.
2. List three characteristics of biological catalysts not shown by other catalysts.

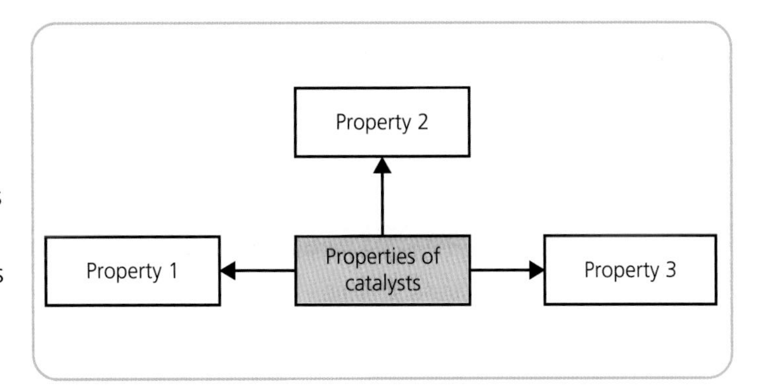

Figure 3.1

Specificity of Enzymes for their Substrate

The stages in Figure 3.2 show enzyme specificity.

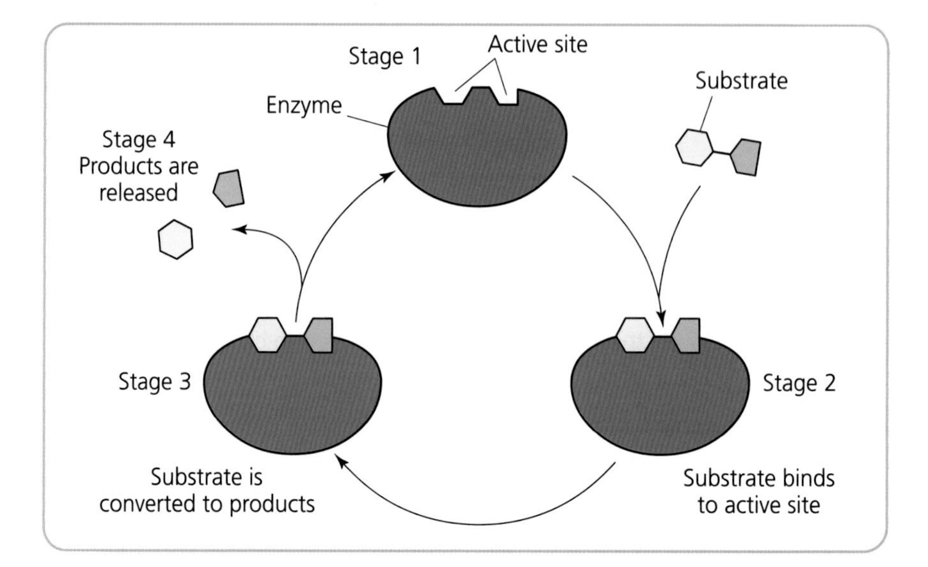

Figure 3.2

Key Points

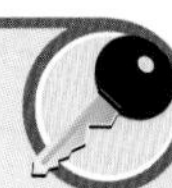

- The chemical that an enzyme reacts with is the **substrate**.
- The area of the enzyme that the substrate binds to is the **active site**.
- **Products** are formed as a result of enzyme action.

  Stage 1 – the active site of the enzyme is free.

  Stage 2 – a substrate molecule binds to the active site.

  Stage 3 – the substrate is converted to products.

  Stage 4 – the products are released from the active site.
- Enzymes are substrate specific.
- There is only one substrate with the shape to bind to the active site of an enzyme.

Stop Think Learn

1. Show the four stages in enzyme specificity as a flowchart.
2. Show the shape of the active sites for substrate molecules A, B and C in figure 3.3.
3. Explain why the enzyme that binds with substrate A is described as being substrate specific.

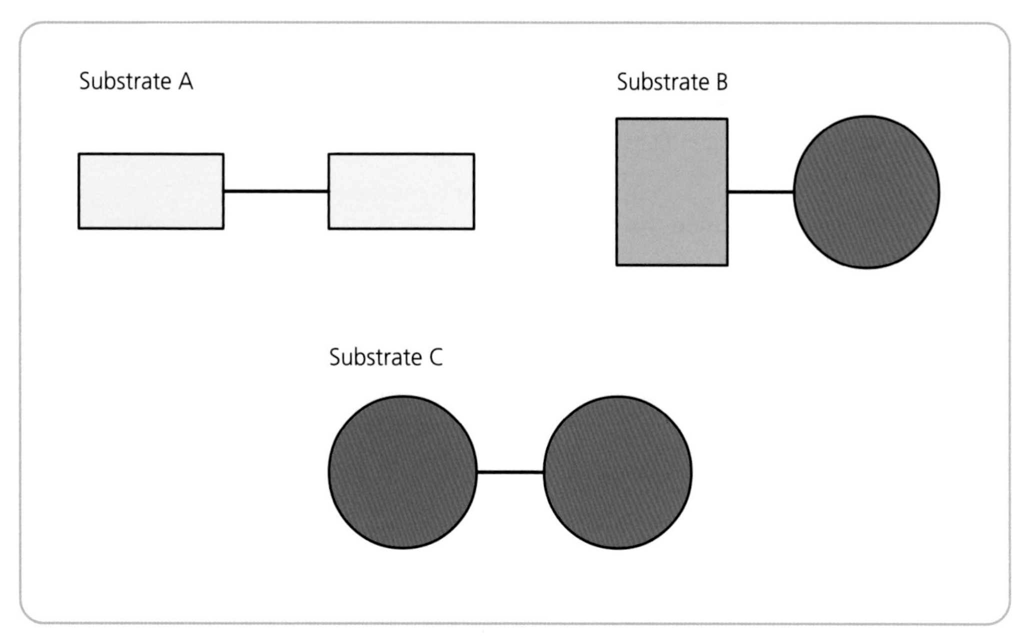

Figure 3.3 **Substrate A, substrate B, substrate C.**

Enzymes Involved in Degradation and Synthesis

Key Points

- Enzymes carry out all chemical reactions in living cells.
- In **degradation**, the bonds holding molecules together are broken.
- The molecules formed as a result of degradation are the product.
- **Synthesis** is the build up of more complex molecules from simpler molecules.
- In synthesis, molecules are bonded together.
- Many substrate molecules are needed to make a complex molecule.

Examples

Examples of degradation

1. Breakdown of starch (takes place in the mouth and small intestine)

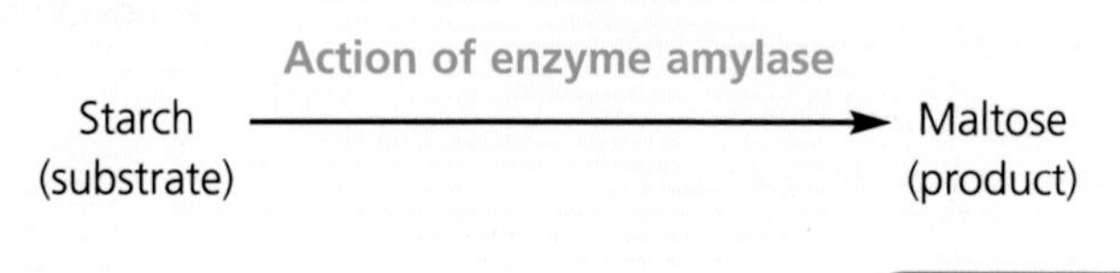

***Examples** continued* ➢

Examples continued

2. Breakdown of hydrogen peroxide (takes place in living plant and animal cells)

Action of enzyme catalase

Hydrogen peroxide (substrate) ⟶ water + oxygen (products)

Example of synthesis

Synthesis of starch (takes place in cells of leaves and in plant storage organs)

Action of enzyme phosphorylase

Glucose-1-phosphate (substrate) ⟶ Starch (product)

Stop Think Learn

1 Explain why degradation and synthesis are opposing reactions.

2 Copy and complete this table to show examples of degradation and synthesis.

Name of enzyme	Substrate	Product(s)	Type of reaction
	Starch		
Catalase			
		Starch	

Factors Affecting Enzyme Activity

Key Points

Figure 3.4 shows the effect of temperature on enzyme activity.

- As temperature increases the reaction rate of the enzyme increases up to 37°C.
- The temperature at which the enzyme operates fastest is the **optimum temperature** – 37°C (see Figure 3.4).
- With further increases in temperature beyond the optimum, the reaction rate decreases and finally the reaction stops.
- Enzymes are proteins and the structure of protein alters at high temperature.

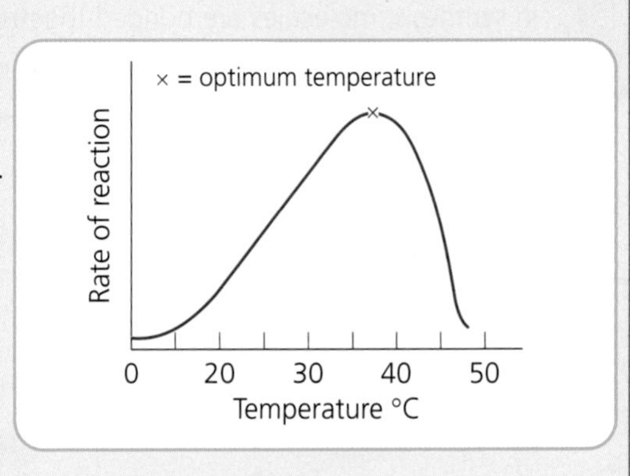

Figure 3.4

Key Points continued ➢

Key Points continued

- Alteration in protein structure changes the shape of the active site.
- As the shape of the active site changes, the substrate can no longer bind to the enzyme and this leads to **inactivation** of the enzyme.
- When the structure of a protein is changed it is **denatured**.

Figure 3.5 shows the effect of pH on enzyme activity.

- Enzymes are active over a range of pH.
- The pH at which the enzyme operates fastest is the **optimum pH**.
- Protein structure is altered with change in pH.
- Alteration in protein structure changes the shape of the active site.

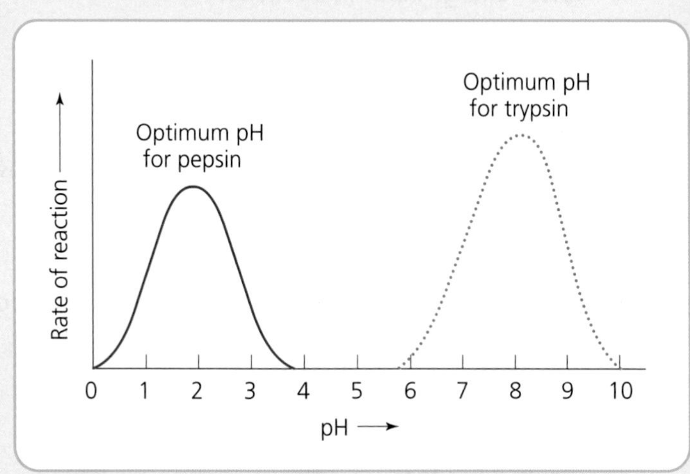

Figure 3.5

- As the shape of the active site changes, the substrate can no longer bind to the enzyme and this leads to inactivation of the enzyme.

Stop Think Learn

1. Name two conditions that affect enzyme reaction.
2. What is meant by the optimum operating conditions for an enzyme?
3. Explain why denaturing an enzyme leads to inactivation of the enzyme.

Hints and Tips

Knowledge of enzymes is needed for the 'Breakdown of food' in Unit 3.

Enzymes are used in many experimental situation questions. Look for these in past papers.

Make up 'flash cards' for the words/phrases that are in **bold** in the text.

Questions

Examination style questions

1 Enzymes are catalysts because they:
  A raise energy output in a reaction.
  B are proteins in chemical nature.
  C lower energy input in a reaction.
  D act on many different substrates.

2 Which is true of the active site?
  A All substrate molecules bind to it.
  B Its shape remains unchanged at high temperatures.
  C Only one type of substrate binds to it.
  D Only one type of product binds to it.

3 Figure 3.6 shows the reaction of an enzyme, a biological catalyst, at a temperature of 30°C.

  a) One of the properties of catalysts shown in Figure 3.6 is that 'catalysts take part in the reaction but remain unchanged'.

State two other properties of catalysts. *(2)*

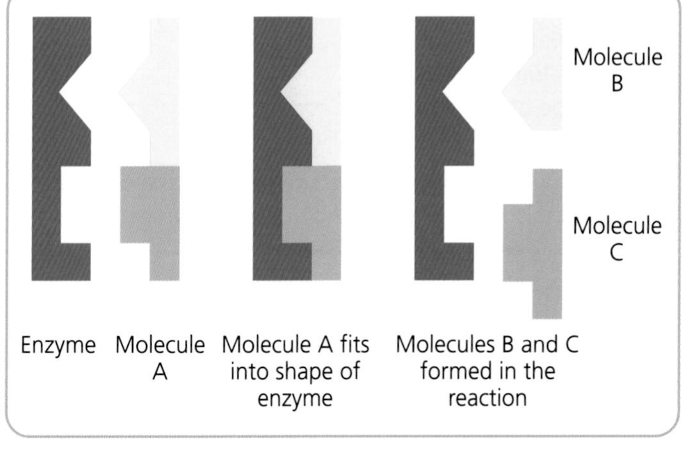

Figure 3.6

  b) Copy and complete the sentence by inserting the correct words.

  Enzymes are found in ____________ cells and are made of ____________. *(1)*

  c) In enzyme reactions, what term is used for molecules such as A? *(1)*

  d) Name the part of the enzyme into which molecule A fits. *(1)*

  e) Figure 3.7 represents the shape of the enzyme after exposure to a temperature of 60°C for 5 minutes.

Use the information to explain why enzymes are inactive after being exposed to high temperatures. *(1)*

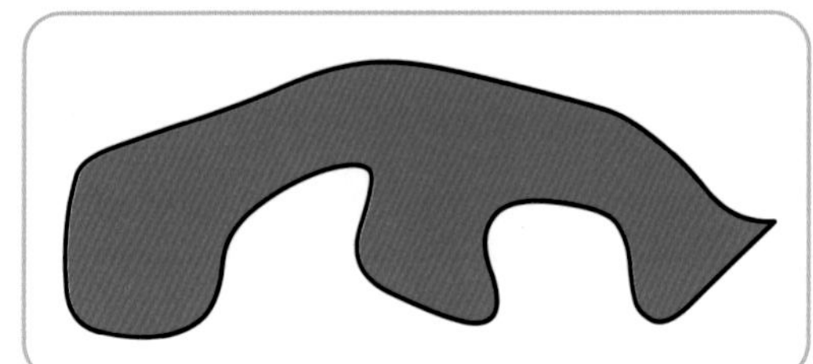

Figure 3.7

Questions continued

4 Figure 3.8 represents a molecule of the enzyme phosphorylase.

Describe and explain the events that take place when the enzyme is added to a solution of glucose-1-phosphate. *(5)*

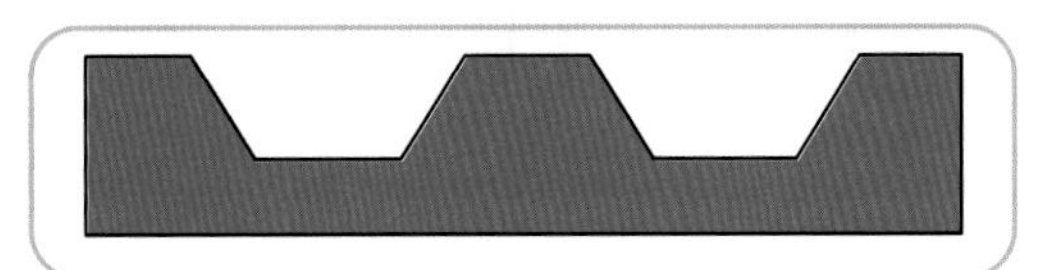

Figure 3.8

Answers

Answers to examination style questions with commentary

1 Enzymes are proteins but **not all** proteins are catalysts. Enzymes act on only **one** substrate. Enzymes lower the energy level. **Answer = C**

2 You should know that the active site is the area on the enzyme to which the substrate binds and that enzymes bind to only one substrate. **Answer = C**

3 a) The list includes:

(i) Lower energy required to cause a chemical reaction.

(ii) Catalysts speed up the rate of chemical reactions.

(iii) Catalysts remain unchanged after taking part in a chemical reaction.

(iv) Catalysts carry out the same reaction many times. **Any two correct.**

b) ALL LIVING and PROTEINS.

c) Molecule A is the **substrate molecule** as it binds to the enzyme.

d) Substrates bind to the **active site**.

e) You should have noticed the change in shape of the enzyme.

Answer = The shape of the enzyme has changed and the substrate molecule can no longer bind to the active site.

4 In extended writing questions you **must** identify all the parts that make up the question. For this question these are:

- Describe the events that take place when the enzyme is added.
- Explain the events that take place when the enzyme is added.

Marking instructions

A1 Glucose-1-phosphate molecules bind to the enzyme.

A2 Glucose-1-phosphate molecules are bonded together.

A3 Starch is formed.

B1 The active site of the enzyme has a specific shape.

B2 Only molecules of glucose-1-phosphate can bind to the active site.

B3 Bonding occurs at the active site.

Each point is valued at 1 mark. **Any 5 correct = maximum total = 5 marks.**

Chapter 4

AEROBIC AND ANAEROBIC RESPIRATION

Glucose as a Source of Energy in Cells

Key Points

- Energy is stored in the bonds of a glucose molecule as **chemical energy**.
- The energy stored in the chemical bonds of a glucose molecule is released in a series of enzyme-controlled reactions.
- The breakdown of glucose to release energy is called **respiration**.
- Some of the energy that is released during respiration is lost as heat energy.
- Most of the energy that is released during respiration is used for **cellular activities**.

  These include:

  (1) muscle contraction,

  (2) cell division,

  (3) synthesis of protein, and

  (4) transmission of nerve impulses.

Stop Think Learn

1 Name the form of energy stored in a glucose molecule.

2 What is respiration?

3 What is the fate of the energy released from the breakdown of glucose that is not used for cellular activities?

4 Copy and complete Figure 4.1 to show four uses of the energy released in respiration.

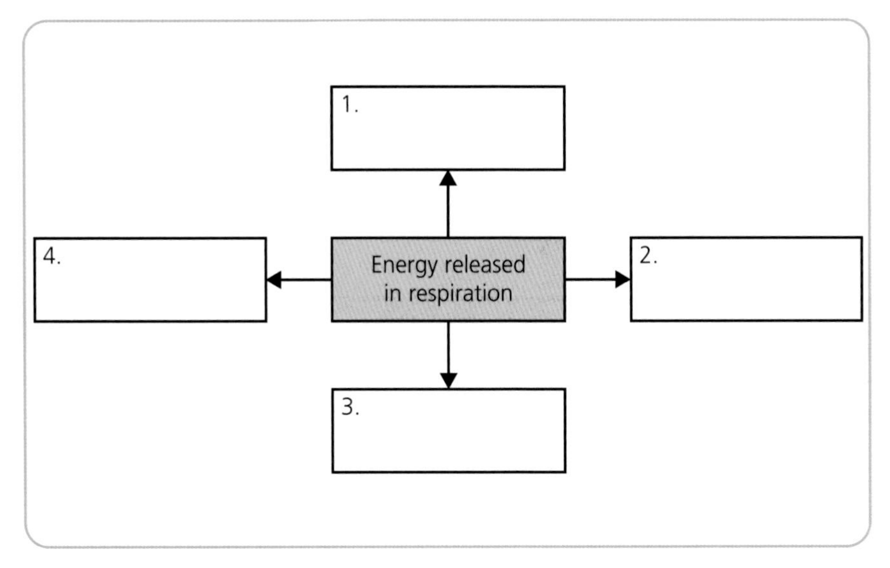

Figure 4.1

Role of ATP

Key Points

- Energy released from the breakdown of a glucose molecule is used to synthesise a chemical molecule called **adenosine triphosphate (ATP)**.
- ATP is synthesised when energy from the breakdown of glucose is used to bond phosphate (Pi) to a molecule of adenosine diphosphate (ADP).
- When ATP breaks down to reform ADP + Pi, energy is released.
- The energy released from the breakdown of ATP is the immediate source of energy for all cellular activities.

The inter-conversion of ADP + Pi and ATP is shown in Figure 4.2.

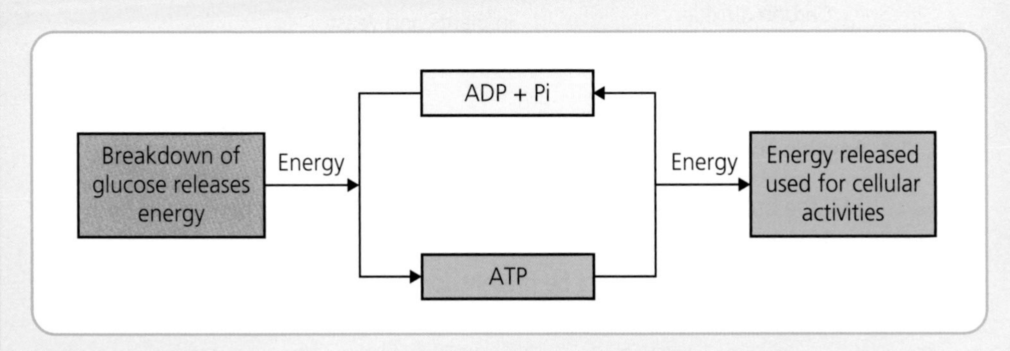

Figure 4.2 ATP cycle

Stop Think Learn

1 Describe the synthesis of ATP.

2 Name the immediate source of energy for all cell activities.

Comparison of Energy Yield and Products of Aerobic and Anaerobic Pathways

Key Points

- **Aerobic respiration** takes place in the presence of oxygen.
- **Anaerobic respiration** takes place in the absence of oxygen.
- Aerobic respiration yields 38 molecules of ATP per glucose molecule.
- Anaerobic respiration yields 2 molecules of ATP per glucose molecule.
- Aerobic respiration is 19 times more efficient than anaerobic respiration.

Figure 4.3 represents an outline of stages in aerobic and anaerobic respiration.

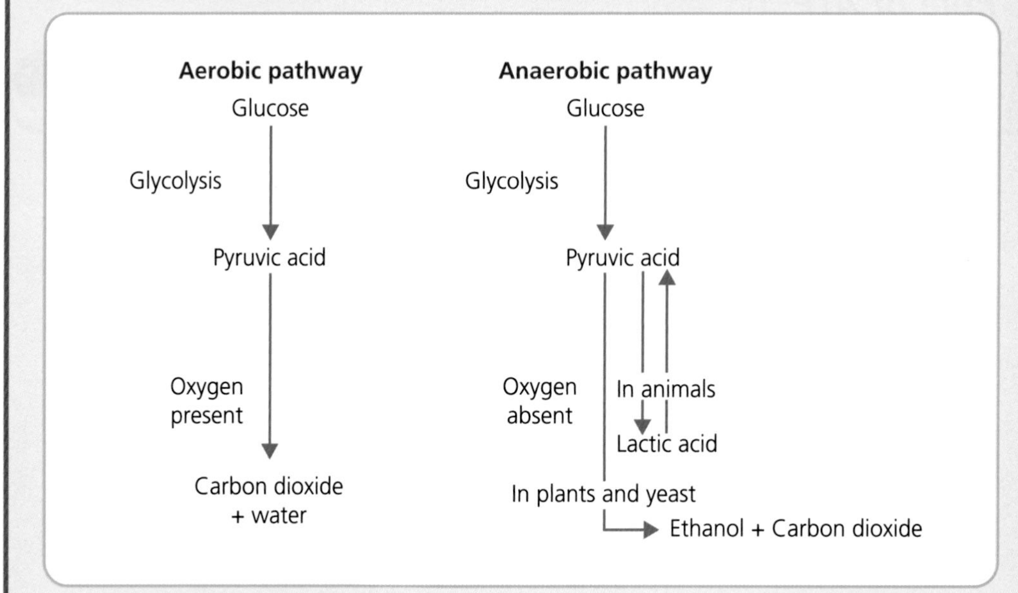

Figure 4.3 Aerobic and anaerobic pathways

- **Glycolysis**, the breakdown of glucose to pyruvic acid, is a stage common to both aerobic and anaerobic respiration.
- In aerobic respiration, oxygen must be present for the complete breakdown of pyruvic acid to carbon dioxide + water.
- When oxygen is not available in the tissues this is called an **oxygen debt** and conditions are anaerobic.

Key Points continued ➢

Key Points continued

- In oxygen debt in animal tissues, pyruvic acid is converted to lactic acid.
- When oxygen becomes available in the tissues the oxygen debt is paid back.
- When oxygen becomes available lactic acid is converted back to pyruvic acid.
- In animal tissues, conversion of pyruvic acid to lactic acid **is reversible**.
- An increase in lactic acid leads to **muscle fatigue** (muscles are less efficient).
- In anaerobic conditions in plant tissue and yeast, the pyruvic acid is broken down to ethanol (alcohol) and carbon dioxide.
- Anaerobic respiration in plant tissue and yeast is **irreversible**.
- Anaerobic respiration is also referred to as fermentation.

Hints and Tips

A**n**aerobic has a letter **N**. N = **NO** = no oxygen present.

In experimental situations, if a gas is given off it is carbon dioxide.

In experimental situations, if the temperature increases it is because of the release of heat energy in respiration.

Stop Think Learn

1 Copy and complete the following word equations for respiration.

a) Aerobic respiration
Glucose + ______________ → Water + ___________ +ATP

b) Anaerobic respiration in muscle tissue
Glucose → ____________________ + ATP

c) Anaerobic respiration in plant tissue and yeast
Glucose → _________________ + _________________ + ATP

2 What is the effect of an oxygen debt in muscle tissue?

3 What is the difference in the number of ATPs produced per glucose molecule in aerobic respiration compared with anaerobic respiration?

4 From the two respiration flowcharts, construct a single flowchart to include both aerobic and anaerobic respiration.

Questions

Examination style questions

1 Which of the following statements describe anaerobic respiration in animal tissue?

1 ATP is synthesised.

2 Lactic acid is produced.

3 Carbon dioxide is produced.

4 Alcohol is produced.

A 1 and 2 only

B 2 and 3 only

C 1, 2 and 3 only

D 1, 3 and 4 only

2 Which of the following is true of the difference between the yield of ATP per glucose molecule in aerobic and anaerobic respiration? The yield is:

A 36 times greater in aerobic respiration.

B 19 times greater in aerobic respiration.

C 36 times greater in anaerobic respiration.

D 19 times greater in anaerobic respiration.

3 Copy and complete this table which refers to a comparison of anaerobic respiration in animal and plant tissue.

Factors relating to anaerobic respiration	Animal tissue	Plant tissue
Number of ATPs produced per glucose molecule	2 ATPs	
Products of conversion of pyruvic acid		Ethanol and carbon dioxide
Conversion of pyruvic acid reversible or non-reversible		

(2)

4 Figure 4.4 represents the synthesis and breakdown of ATP.

a) Name substances X and Y. *(1)*

b) Give an example of a cellular activity that requires energy from ATP. *(1)*

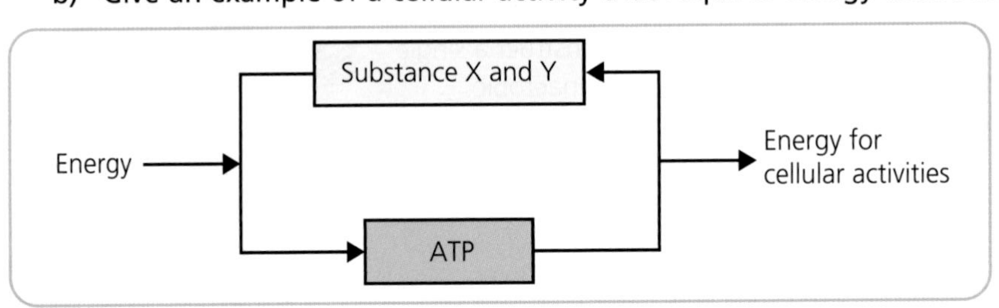

Figure 4.4

Questions continued

5 Figure 4.5 represents an outline of stages in aerobic and anaerobic respiration.

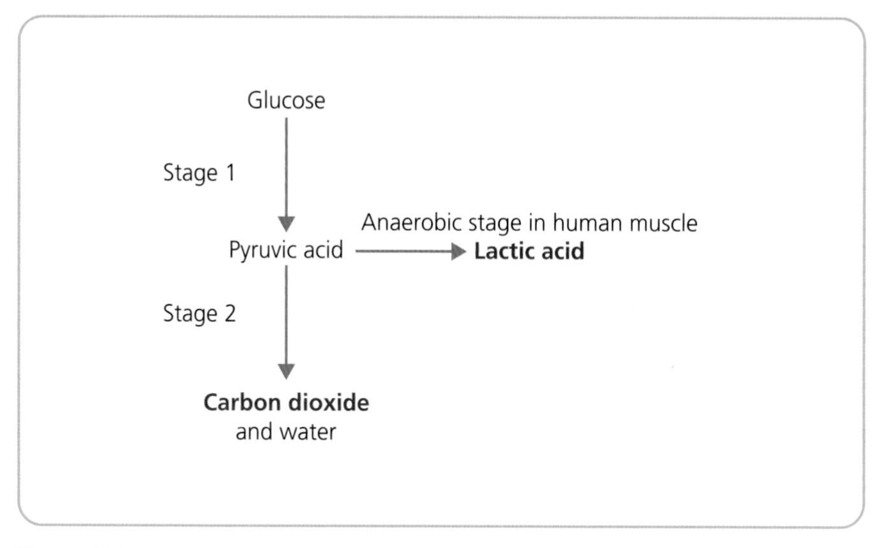

Figure 4.5

a) Which of the stages proceeds only when oxygen is present? *(1)*

b) Name stage 1. *(1)*

c) State the effect of a build up of lactic acid on muscle tissue. *(1)*

d) The build up of lactic acid is because of an oxygen debt. What happens to the lactic acid as the oxygen debt is paid back? *(1)*

e) In what form is energy lost during respiration? *(1)*

6 Describe the process of aerobic respiration. *(5)*

Answers

Answers for examination style questions with commentary

1 ATP is synthesised in anaerobic respiration; lactic acid is produced in animal tissue; alcohol and carbon dioxide are not produced. **Answer = A**

2 Aerobic respiration produces 38 ATP molecules and anaerobic respiration produces 2 ATP molecules. This works out as being 19 times greater in aerobic respiration. **Answer = B**

3 Glycolysis is the common stage to both plant and animal anaerobic respiration. Therefore, **2 ATPs** in plant tissue.

The product of anaerobic respiration from pyruvic acid in animal tissue is **lactic acid**.

Pyruvic acid to lactic acid is **reversible**. Pyruvic acid to ethanol and carbon dioxide is **irreversible**.

4 correct = 2 marks; 3/2 correct = 1 mark.

Answers continued

4 a) X and Y are ADP and Pi. Both for 1 mark.

b) Muscle contraction or cell division or synthesis of protein or transmission of nerve impulse.

5 a) Oxygen is required for the aerobic stage. This is shown by the conversion of pyruvic acid into carbon dioxide and water so **answer = stage 2.**

b) Conversion of glucose to pyruvic acid is **glycolysis**.

c) Build up of lactic acid leads to **muscle fatigue**.

d) This is testing if you know that conversion is reversible.

Answer = Lactic acid is converted to pyruvic acid.

e) You should know that in energy conversions energy is lost as **heat energy**.

6 Describe the process of aerobic respiration.

In extended writing questions you **must** identify all the parts that make up the question. These are:

- Definition of aerobic respiration.
- Describe glycolysis.
- Describe the oxygen-requiring stage.
- Describe how ATP is formed.

Marking Instructions

A1 Respiration is the process by which cells release energy.

A2 Respiration involves a series of enzyme-controlled reactions.

A3 Energy is used to produce ATP from ADP + Pi.

B1 Glucose is broken down to pyruvic acid.

B2 This first stage is called glycolysis.

B3 Two molecules of ATP are produced in this stage.

C1 The pyruvic is broken down to carbon dioxide and water.

C2 Oxygen must be present for this reaction.

C3 Thirty-eight ATP molecules are produced from one glucose molecule.

No more than 2 marks from each group A, B and C to a maximum of 5 marks.

Chapter 5

PHOTOSYNTHESIS

Sunlight as the Source of Energy

Key Points

Figure 5.1 shows the structure of a palisade leaf cell and the path of gas exchange in the light.

- The **chloroplast** contains chlorophyll.
- **Chlorophyll** absorbs light energy from the sun.
- Within the chloroplast, light **energy is converted** to chemical energy.
- Light energy supplies the energy to bond Pi to ADP to **make ATP**.
- **ATP is used** in the production of glucose.
- Carbon dioxide diffuses into the leaf cell.
- Oxygen diffuses out from the leaf cell.

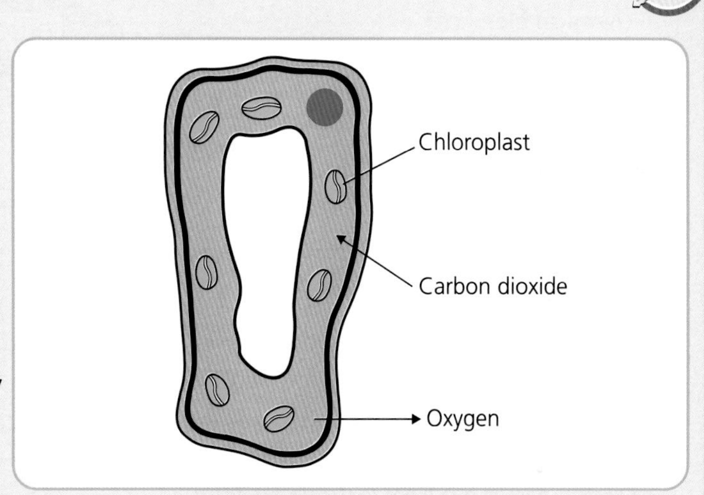

Figure 5.1

Equation for Photosynthesis

Key Points

Figure 5.2 outlines the process of photosynthesis.

- The **product of photosynthesis** is the carbohydrate glucose.
- **Glucose** contains the **chemical elements** carbon, hydrogen and oxygen.
- The plant must have a source of **raw materials** that supply carbon, hydrogen and oxygen for the synthesis of glucose.

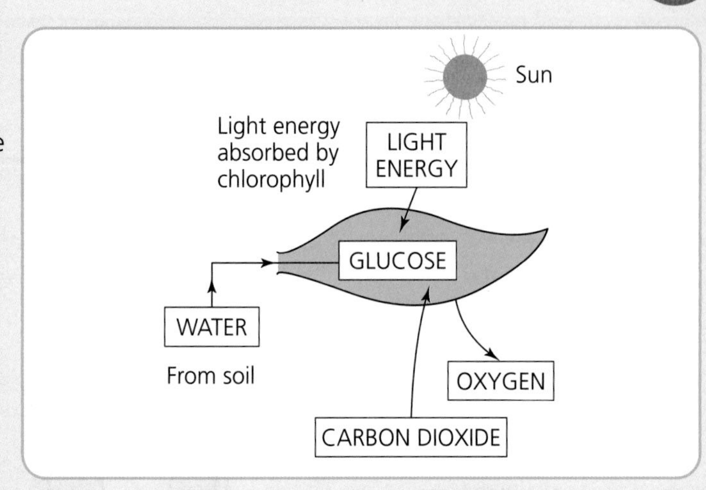

Figure 5.2

- The **source of carbon and oxygen** is carbon dioxide.
- The **source of hydrogen** is water and the oxygen of the water is a by-product.
- **Diffusion is important in photosynthesis** as the raw material carbon dioxide diffuses into the leaf from the surrounding air.
- Oxygen passes out from leaf cells by diffusion.

Word equation for photosynthesis

$$\text{Carbon dioxide + Water} \xrightarrow[\text{by chlorophyll}]{\text{light energy absorbed}} \text{Glucose + oxygen}$$

(Raw materials) (By-product)

Stop Think Learn

Copy and complete Figure 5.3 to show energy transfer in photosynthesis.

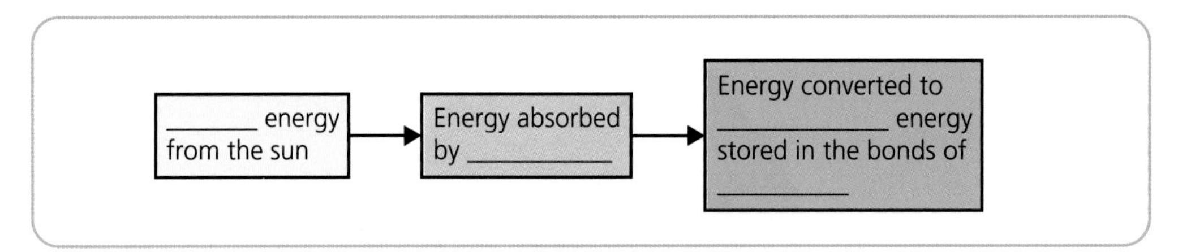

Figure 5.3

Photolysis and Carbon Fixation

Key Points

Stage 1 Photolysis

Figure 5.4 outlines the process of photolysis

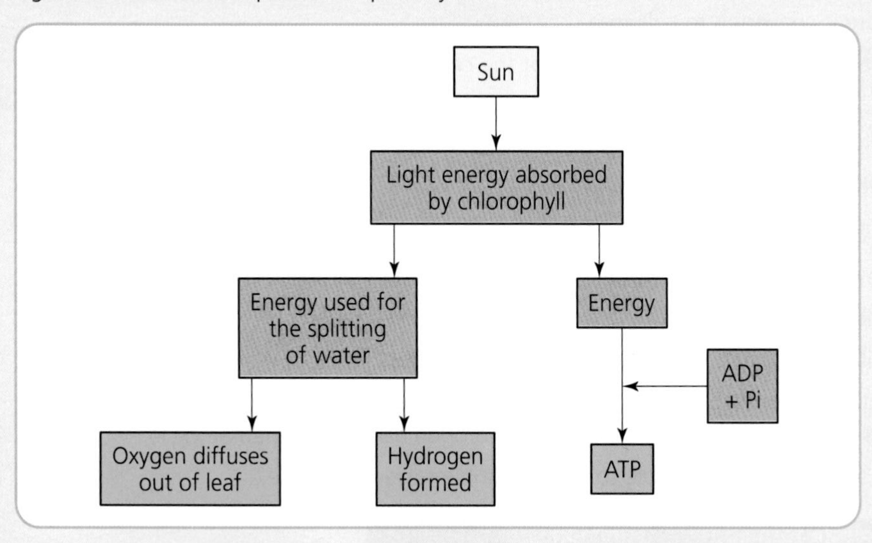

Figure 5.4

- In the process of **photolysis**, energy absorbed by chlorophyll is used to split water into hydrogen and oxygen.
- Oxygen is a **by-product** and diffuses out of the cells and out of the leaf.
- A **hydrogen carrier** picks up hydrogen.
- The carrier transports hydrogen to the carbon fixation stage.
- **Light energy** absorbed is used to bond ADP and Pi to make ATP.
- ATP is used in carbon fixation.

Stage 2 Carbon fixation

Figure 5.5 outlines the process of **carbon fixation**.

- The hydrogen carrier transports hydrogen, produced in photolysis, to the carbon fixation stage.
- **Hydrogen** is combined with carbon dioxide.
- ATP supplies the energy.
- As energy is released ADP + Pi are reformed.
- Glucose is produced as a result of these reactions.

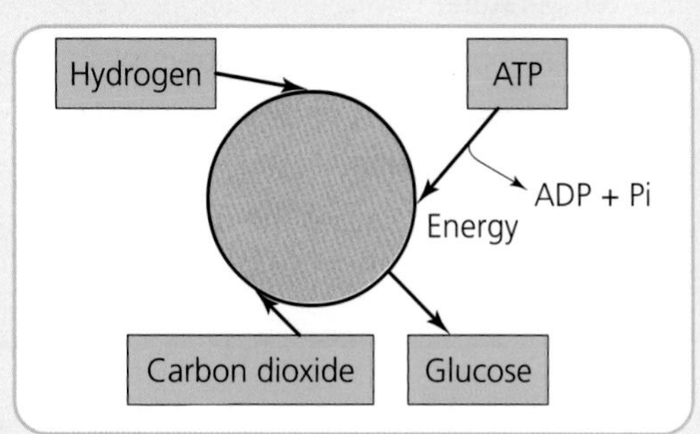

Figure 5.5

Conversion of Glucose to Other Carbohydrates

Key Points

- Glucose is converted to starch.
- **Starch** is a storage carbohydrate in plants, for example, stored in potato tubers.
- Glucose is converted to cellulose.
- **Cellulose** is a structural component of the cell wall of plant cells.

Stop Think Learn

1. Describe the process of photolysis.
2. Name the two products of photolysis that are used in carbon fixation.
3. Describe how hydrogen is transferred to the carbon fixation stage.
4. Describe the formation of glucose in carbon fixation.
5. From the two outlines (figure 5.4 and 5.5), construct a single flowchart to show the stages of photosynthesis to include the hydrogen carrier and the ATP cycle.
6. Copy and complete the table to show possible fates of glucose.

Carbohydrate formed	Function
	Storage carbohydrate
Cellulose	

Limiting Factors

Key Points

- The rate of photosynthesis depends on light intensity, carbon dioxide concentration and temperature.
- For the optimal rate of photosynthesis, all three factors must be at their optimal level.
- If any one factor is under its optimal level, the rate of photosynthesis will be less than the expected optimal rate.
- The factor that causes the decrease in the rate of photosynthesis is called the **limiting factor**.

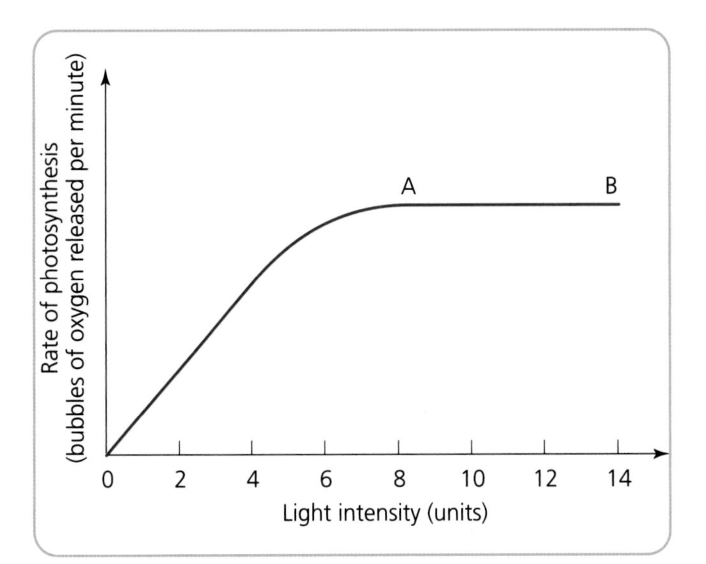

Figure 5.6 shows the effect of light intensity on the rate of photosynthesis when temperature and carbon dioxide concentration remain constant.

Figure 5.6

Figure 5.7 shows the effect of carbon dioxide concentration on the rate of photosynthesis when temperature and light intensity remain constant.

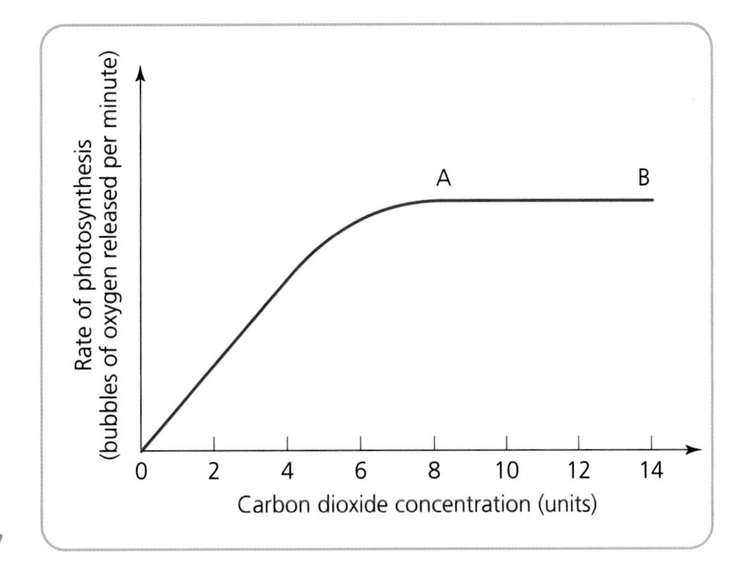

Figure 5.7

In Figures 5.6 and 5.7, from point 0 to A, as light intensity and carbon dioxide concentration increase the rate of photosynthesis increases.

In Figure 5.6, the factor that prevents the rate of photosynthesis reaching its optimal level between point 0 and A is light intensity. Light intensity is the limiting factor.

In Figure 5.7, the factor that prevents the rate of photosynthesis reaching its optimal level between point 0 and A is carbon dioxide concentration. Carbon dioxide concentration is the limiting factor.

In Figure 5.6 and 5.7, from point A to B, further increases in light intensity and carbon dioxide concentration do not increase the rate of photosynthesis.

In Figures 5.6 and 5.7, from point A to B, it must be one of the other factors that prevent photosynthesis reaching the optimal rate.

Production of Early Crops in Horticulture

Key Points

- Horticulturists grow crops under artificial conditions in large greenhouses or under polythene frames. From their knowledge of limiting factors, they ensure that conditions for the production of the crop are at their optimum.
- Supplementary lighting is used to ensure optimum light intensity.
- Carbon dioxide is pumped into the atmosphere to ensure optimum levels.
- Heating is used to ensure the optimum temperature.

Hints and Tips

Always, when referring to light, use the terms light intensity and light energy.

Glucose produced in photosynthesis is used as the energy source in respiration for the plant. DO NOT FORGET that plants respire at all times.

In graphs on limiting factors, as long as the curve has not levelled off, the limiting factor is that shown on the x-axis.

When the graph levels off one of the other two factors is the limiting factor.

Rate involves time. When asked about rate of photosynthesis, you must use time or units of time in your answer.

In many experimental situations, you must realise that measurements such as the number of bubbles of gas per minute are used to measure the rate of photosynthesis.

Temperature for photosynthesis follows the same pathway as the effect of temperature on enzyme action. Proteins/enzymes are denatured at high temperatures.

Make 'flash cards' for the words and phrases in **bold** in the text.

Stop Think Learn

1 Name three factors that affect the rate of photosynthesis.
2 Explain what is meant by a limiting factor.
3 Copy and complete Figure 5.8 to show how horticulturists ensure optimal conditions for crop growth.

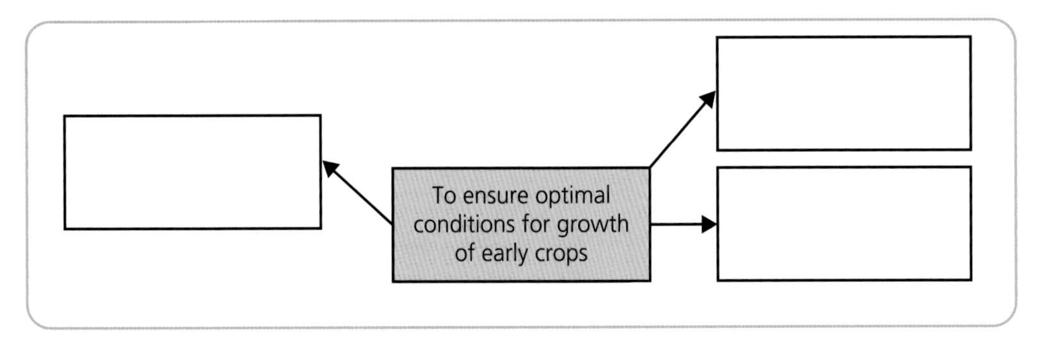

Figure 5.8

Questions

Examination style questions

1 The set up in Figure 5.9 was used in an investigation.

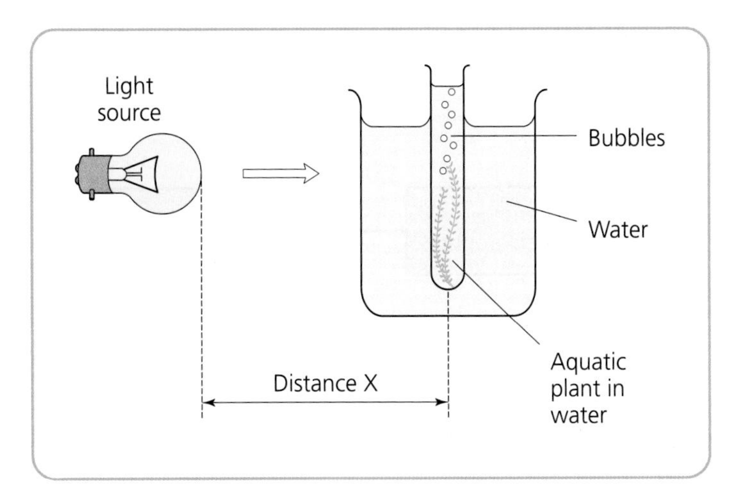

Figure 5.9

When distance X is increased, which of the following remains unchanged?

A light intensity at the plant

B rate of bubble production

C rate of respiration of the plant

D rate of photosynthesis of the plant

Questions continued

2 Which of the following is both a raw material and a limiting factor for photosynthesis?

A glucose

B carbon dioxide

C light energy

D water

3 Photolysis is the:

A energy release from splitting water.

B splitting of water by light energy.

C combining of hydrogen with carbon dioxide.

D synthesis of ATP from ADP + Pi.

4 The diagram shows a summary of a stage in photosynthesis that takes place in the chloroplasts of leaf cells.

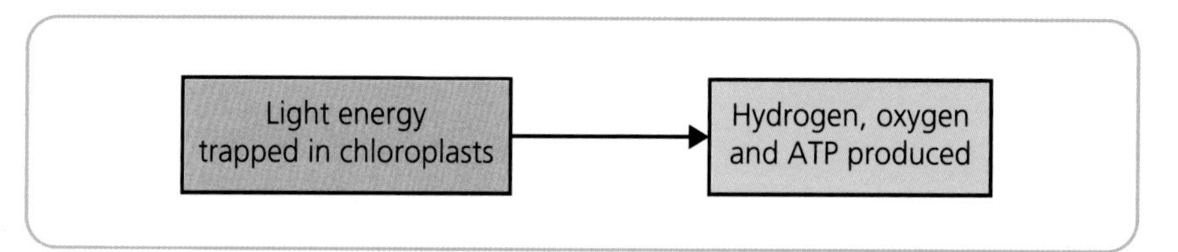

Figure 5.10

a) Name the chemical present in the chloroplast that traps light energy. *(1)*

b) Describe how hydrogen and oxygen are produced during this stage of photosynthesis. *(1)*

c) Describe the energy conversion that takes place in the production of ATP. *(1)*

d) What use does the cell make of the ATP produced in this stage? *(1)*

e) Name the process by which oxygen moves out of the leaf cells. *(1)*

5 Figure 5.11 shows the effects of increasing carbon dioxide concentration on the rate of photosynthesis at different light intensities and temperatures.

Questions continued

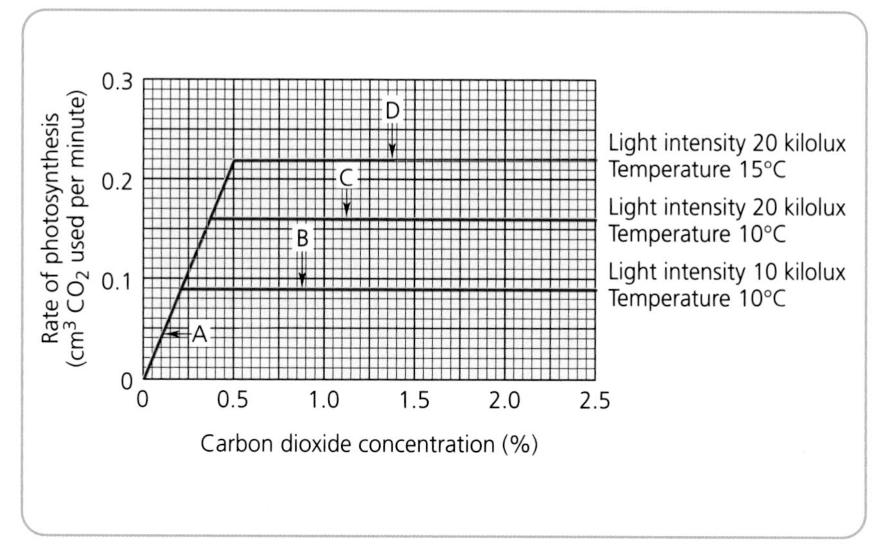

Figure 5.11

a) **Identify the factor that limits the rate of photosynthesis at points A and C.** *(1+1)*

b) **From Figure 5.11, calculate the difference in the rates of photosynthesis between points B and D.** *(1)*

c) **How do horticulturists make use of the results from such investigations?** *(1)*

6 **Describe the absorption of light energy and its use in photolysis.** *(5)*

Answers

Answers to examination style questions with commentary

1 Moving the lamp away decreases light intensity, rate of bubbles and photosynthesis. Respiration is not affected by light intensity. **Answer = C**

2 Water and carbon dioxide are raw materials but of these carbon dioxide is the only one that is a limiting factor. **Answer = B**

3 You should know that photolysis is the splitting of water using light energy absorbed by chlorophyll. **Answer = B**

4 a) The chemical in chloroplast that traps light energy is **chlorophyll**.

b) Hydrogen and oxygen are produced when **energy is used to split water**.

c) In ATP production, light energy is converted to chemical energy.

Answers continued

d) The cell uses ATP to make glucose in carbon fixation.

e) Oxygen moves out of the leaf cells by diffusion.

5 a) At point A, the rate of photosynthesis is still increasing with increasing carbon dioxide concentration. The limiting factor is **carbon dioxide concentration**. At point C, light intensity is 20 kilolux and temperature is 10°C. At point D, where the rate of photosynthesis is greater, the light intensity is the same at 20 kilolux but the temperature is at 15°C. The limiting factor is **temperature**.

b) Each small square = 0.01 units. At B value = 0.09 units and at D = 0.22 units. Difference is 0.22 – 0.09 = **0.13 cm^3 CO_2 used per minute**.

c) They ensure that conditions for light intensity, carbon dioxide concentration and temperature will be at the optimal levels for growth.

6 Describe the absorption of light energy and its use in photolysis.

In extended writing questions you **must** identify all the parts that make up the question. These are:

- absorption of light energy,
- use of energy in photolysis.

Marking instructions

A1 The source of light energy is the sun.

A2 Light energy is absorbed by chlorophyll.

A3 Light energy is converted to chemical energy.

B2 Energy is used to produce ATP from ADP + Pi.

B3 Energy is used to split water into hydrogen and oxygen.

B4 The splitting of water is called photolysis.

B5 Oxygen is a by-product.

B6 Hydrogen is picked up by a hydrogen carrier.

B7 Hydrogen and ATP are used in carbon fixation.

No more than 2 marks from group A, and no more than 4 marks from group B, to a maximum of 5 marks.

UNIT 2

Environmental Biology and Genetics

Chapter 6

ECOSYSTEMS

Components of an Ecosystem

Key Points

- Habitats, populations and communities are components of an ecosystem.
- **Habitat** is the place where an organism lives in an ecosystem.
- **Population** is the number of individuals of the same species in an ecosystem.
- **Community** is all the populations of different species in an ecosystem.
- The **niche** of an organism in an ecosystem is shown by what it feeds on and what feeds on it.

Food Chains and Food Webs

Key Points

The feeding relationship between organisms in an ecosystem is shown by **food chains**. Identify the following from the generalised food chain in Figure 6.1:

- **Producers** are green plants.
- Producers make their own food by photosynthesis.
- **Primary consumers** obtain energy by eating plant materials.
- Primary consumers are **herbivores**.
- **Secondary consumers** obtain energy by eating other animals.
- Secondary consumers are **carnivores**.
- **Predators** feed off other animals (carnivores).
- **Prey** are the animals eaten by predators.
- **Decomposers** obtain energy by feeding off the dead remains, faeces, fallen leaves etc. of other organisms.
- Activity of decomposers releases nutrients into the soil.
- Decomposers are bacteria and fungi.

Not shown in the food chain are **omnivores**.

- Omnivores obtain energy by eating both plant material and other animals.

***Key Points** continued* ➢

Key Points continued

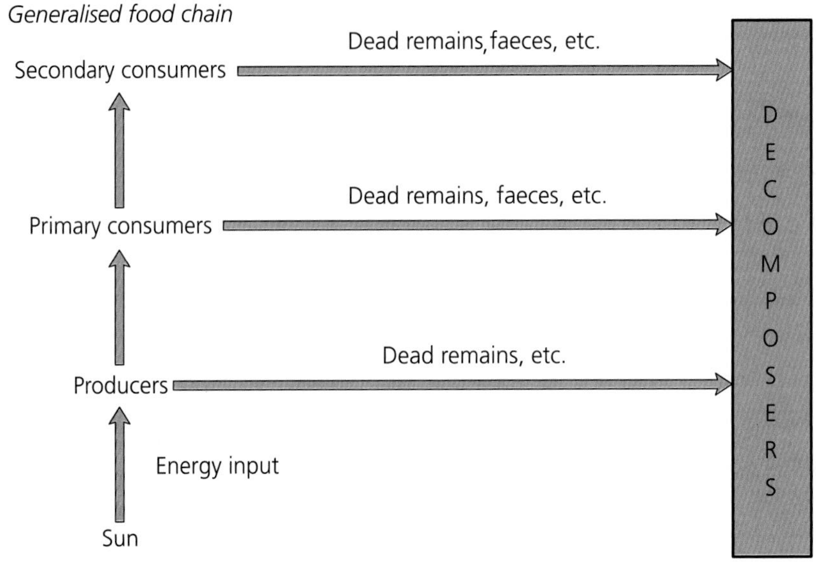

Generalised food chain

Secondary consumers → Dead remains, faeces, etc. → DECOMPOSERS

Primary consumers → Dead remains, faeces, etc. → DECOMPOSERS

Producers → Dead remains, etc. → DECOMPOSERS

Energy input

Sun

Figure 6.1

- **Arrows** show the direction of energy transfer between organisms.
- A more complete picture of the feeding (energy transfer) relationships in an ecosystem is shown in a **food web**. The food web links all the food chains in an ecosystem. Figure 6.2 shows part of a food web in a Scottish ecosystem.

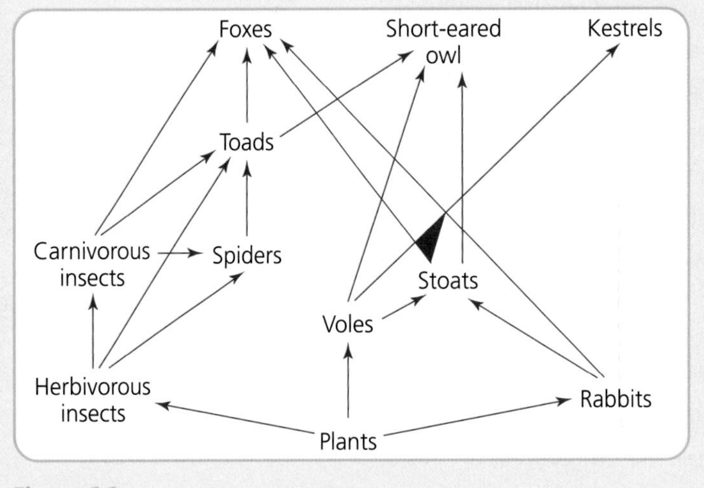

Figure 6.2

Stop Think Learn

1. From the food web (see Figure 6.2) identify the following:
   a) producers,
   b) primary consumers,
   c) secondary consumers only,
   d) all carnivores other than secondary consumers,
   e) predators,
   f) prey.
2. Show the position of sparrows in the food web, given that sparrows eat plants, insects and spiders and that they are preyed upon by kestrels.
3. What term describes feeding of the sparrow?

Flow of Energy Through Food Chains in Ecosystems

Key Points

- On average, 90% of the energy transferred from one feeding level to the next is lost.
- **Energy is lost** mainly as heat energy during respiration and in animal faeces.
- The energy in faeces is lost in the sense that it has not been built into the tissue of the animal. It is not lost in that it is an energy source for decomposers.

Figure 6.3 shows energy flow in a food chain that starts from 100 units of energy of plant material.

Thus of the original 100 units of energy in the plant material, 1 unit only is built into the tissue of the kestrel.

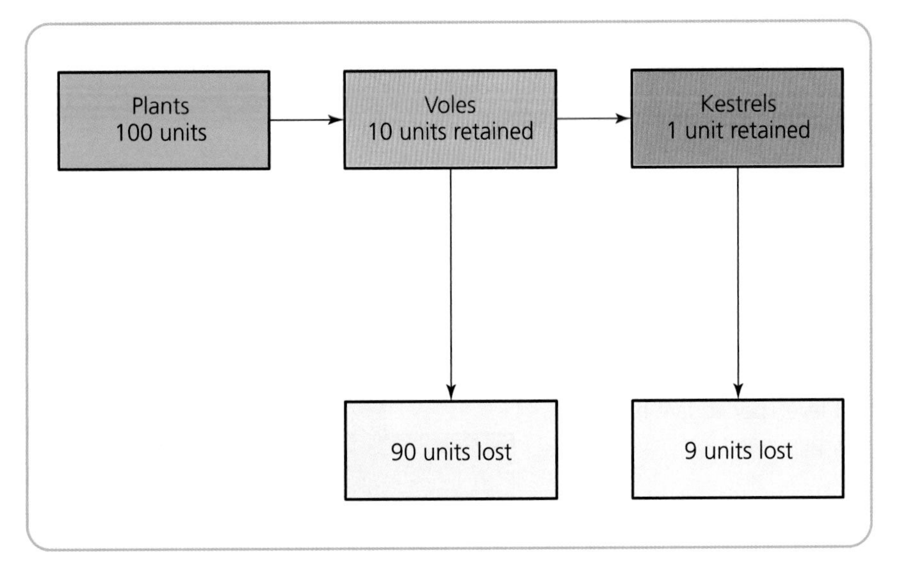

Figure 6.3

Pyramid of numbers

These are used to represent populations within a food chain.

The **pyramid of numbers** is a diagram that shows population numbers at each feeding level in a food chain.

In general, the numbers decrease on moving along the food chain (Figure 6.4).

The weakness of a pyramid of numbers is with a large producer such as an oak tree. (Figure 6.5). A single oak tree can be a source of food for thousands of insects.

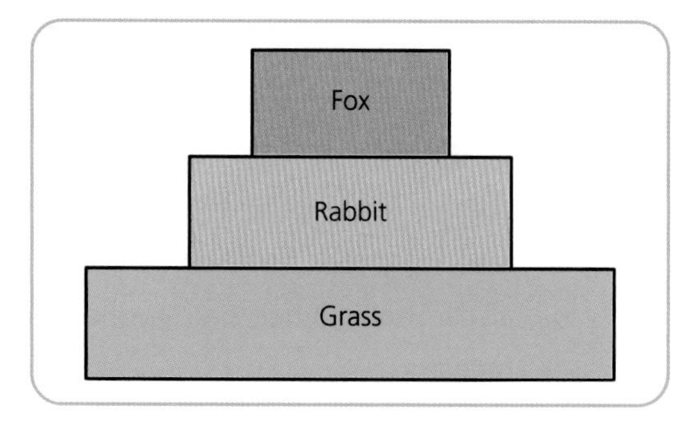

Figure 6.4

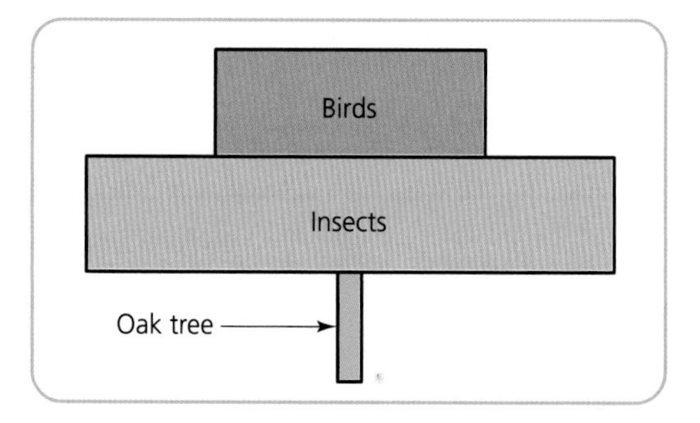

Figure 6.5

Pyramid of biomass

A **pyramid of biomass** is used to overcome the weakness in the pyramid of numbers.

The pyramid of biomass is a diagram that shows the biomass of a population in an ecosystem (Figure 6.6).

Biomass is measured as total dry mass.

Biomass is measured at a particular point in time.

Biomass of a population in an ecosystem varies throughout the year.

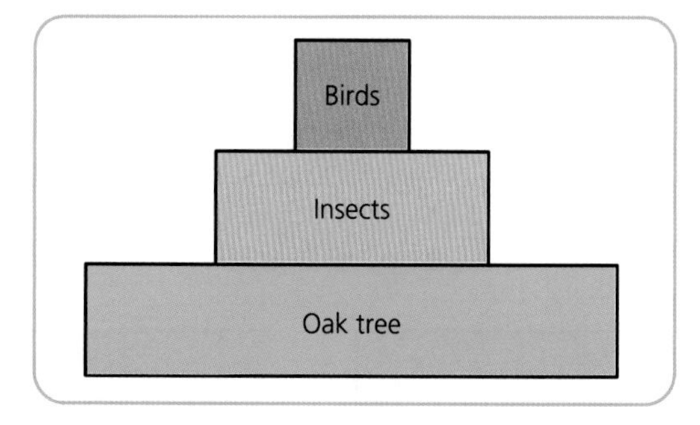

Figure 6.6

Pyramid of energy

These are used to overcome the weakness of the variability in biomass throughout the year.

A **pyramid of energy** is a diagram that shows the flow of energy through an ecosystem (Figure 6.7).

Each layer of the pyramid represents a feeding level.

The length of the layer is proportional to the total energy flowing through the organisms in that level, per square metre of the ecosystem, per year.

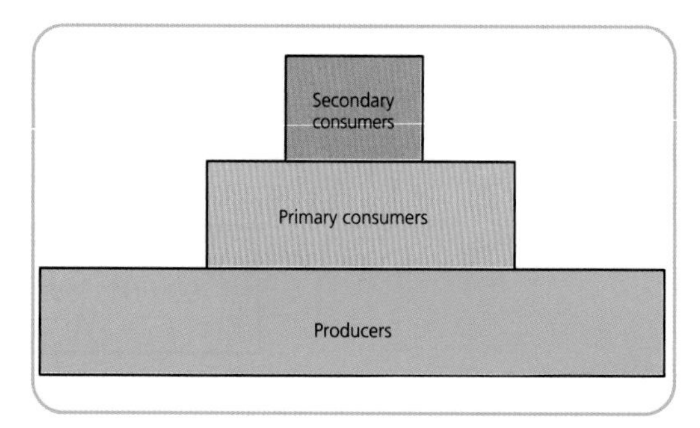

Figure 6.7

The Importance of Biodiversity at Species Level

Key Points

Biodiversity is the number of different species in an ecosystem.

A **species** is a group of organisms that interbreed to produce fertile offspring.

Stable ecosystems have large numbers of different species in their food webs.

Consequences of removal of a species from a food web

Removal of a species from a food web is often the result of human activity.

It occurs through over-fishing, over-hunting, disease, climate change etc.

In Figure 6.2, what would be the consequences on the population of rabbits if all the foxes were hunted? Give a reason for your answer.

The answer has to be that either it increases or decreases or stays the same.

The reason has to be based on the logic in the information.

Possible answer 1: Population increases.

Reason: Removal of the foxes means that fewer rabbits are eaten.

Answer 2: Population stays the same.

Reason: Although rabbits are not eaten by foxes, the stoats eat more rabbits as there is no competition from foxes.

In Figure 6.2, what would be the consequences on the population of herbivorous insects if all the foxes were hunted? Give a reason for your answer.

Answer: Population increases.

Reason: Toads are not eaten by foxes, so there are more toads to eat carnivorous insects and spiders thus fewer herbivorous insects are eaten.

Stop Think Learn

1 State the meaning of the following:
   a) biodiversity,
   b) species.
2 Describe a feature of a stable ecosystem.
3 Looking back at figure 6.2, predict the effect on the population of voles if all the short-eared owls were removed from the food web. Give a reason for your answer.

Factors Affecting Biodiversity

Adaptations to habitat and niche

Darwin's finches show adaptations of beak shape and size.

The Galapagos Islands lie 1500 kilometres from mainland South America.

It is believed that at one time in the past, members of a species of finch were blown across to the islands and evolved in isolation.

Figure 6.8 shows how the **shape and size of the beaks** of species of finch are adapted to exploit different food sources.

The shapes of the beaks differ in width and length.

Wider beaks are for crushing seeds, soft fruit and nuts.

Longer beaks are for capturing insects in flight and for searching into bark for grubs.

The finches adapted to fill niches on the islands that were vacant.

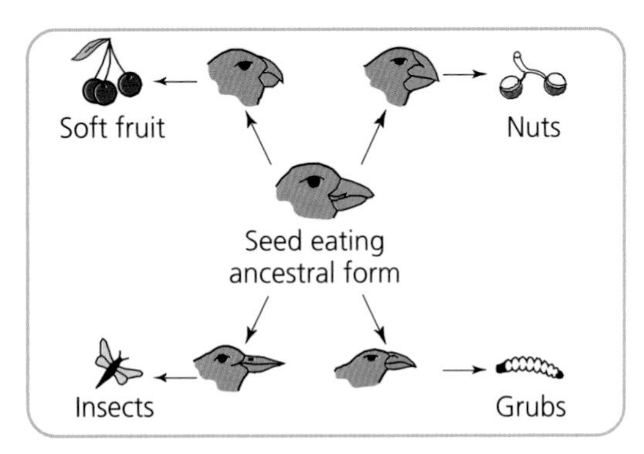

Figure 6.8

The effect of adaptations of beak shape and size allows more niches to be occupied and to increase biodiversity.

Desert plants – adaptations of roots and leaves

The adaptations are in response to low water availability and these either increase water uptake or reduce water loss.

Figure 6.9 shows a cactus plant and a section through one of its leaves.

Deep roots penetrate into the soil and can reach down to the water table.

The **surface roots** spread over a large area. These absorb water that condenses on the surface of the soil during the colder desert nights.

Leaves are shaped like needles. This reduces the surface area and thus decreases water loss by evaporation from the leaves.

The **thick waxy cuticle** gives an impervious cover and this decreases water loss by evaporation from the leaf.

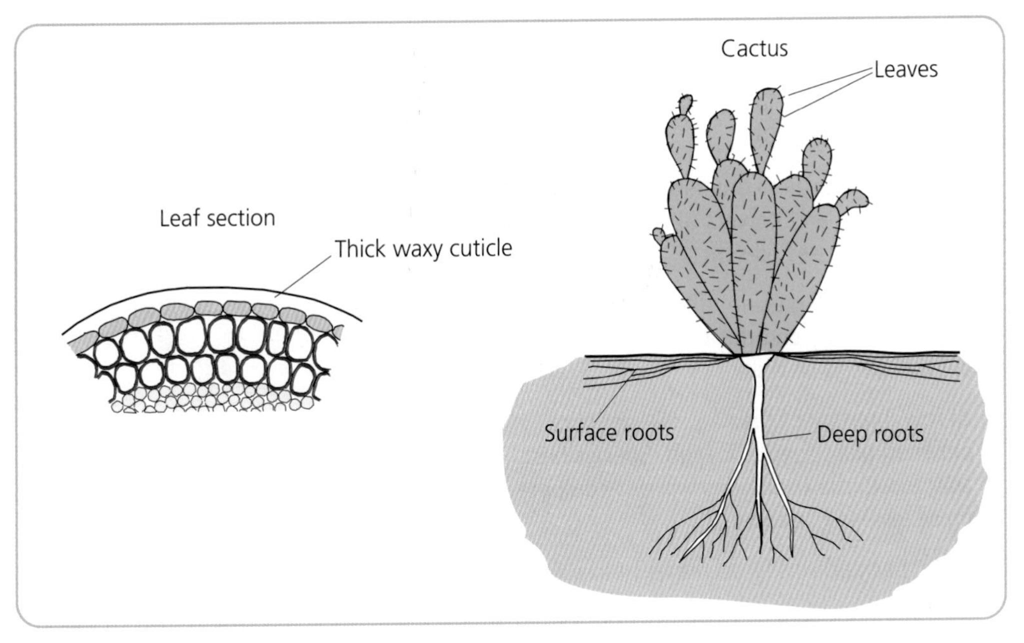

Figure 6.9

Competition in plants and animals and how it affects biodiversity

All species of green plant have to obtain water, light energy and nutrients (mineral salts) from the soil.

Within ecosystems there is **competition** between plants for these factors.

Other factors that affect the growth of a plant within the environment can differ.

Conditions such as type of soil, pH of soil, light availability, etc.

Some plant species will be better adapted than others to the conditions and they will succeed in competition for water, light energy and mineral salts.

A species that competes well in one set of conditions may fail in another.

All animal species compete for food, water and shelter.

From Figure 6.2, stoats, short-eared owls and kestrels are in competition for voles. They are in competition for the same prey.

Stop Think Learn

Complete the table on adaptations of a desert plant.

Plant structure	Adaptations	Benefit to plant

Effects of grazing on biodiversity

The feeding of many herbivores such as deer, goats, sheep, etc. is described as grazing (Figure 6.10).

Figure 6.10

When grazing, these animals crop the leaves, stems and flowers of the various herbs and grasses present.

The intensity of grazing affects plant species diversity and this in turn affects animal species diversity.

At low intensity of grazing, competition between plant species is high. Species best adapted to the conditions will out-compete those less well adapted.

Thus species diversity decreases at low intensity of grazing.

At continuous very high intensity of grazing annual plants fail to develop seeds. These species gradually die out due to reproductive failure.

Thus species diversity decreases at very high intensity of grazing.

At high intensity of grazing, none of the problems associated with low or very high intensity of grazing exist, thus, species diversity is maintained.

Effects of human activity on biodiversity

Sources of **pollution** in an ecosystem can be agricultural, domestic and industrial.

Pollutants include crop sprays, gases from burning of fossil fuels, etc.

Many of the pollutants cause plants to die and thus decrease species diversity.

Habitat destruction in an ecosystem is through deforestation, removal of hedgerows, ploughing of natural grasslands, for planting crops, etc.

Habitat destruction leads to a decrease in species diversity.

Behavioural Adaptations in Animals and their Adaptive Significance

Animals must be able to respond to their environment in order to survive.

For example, many species of woodlice loose water rapidly when in a dry area and are in danger of dying as a result of desiccation.

The apparatus in Figure 6.12 is used to demonstrate the behavioural response of woodlice to relative humidity.

Results from such investigations are shown below.

Figure 6.11 A woodlouse

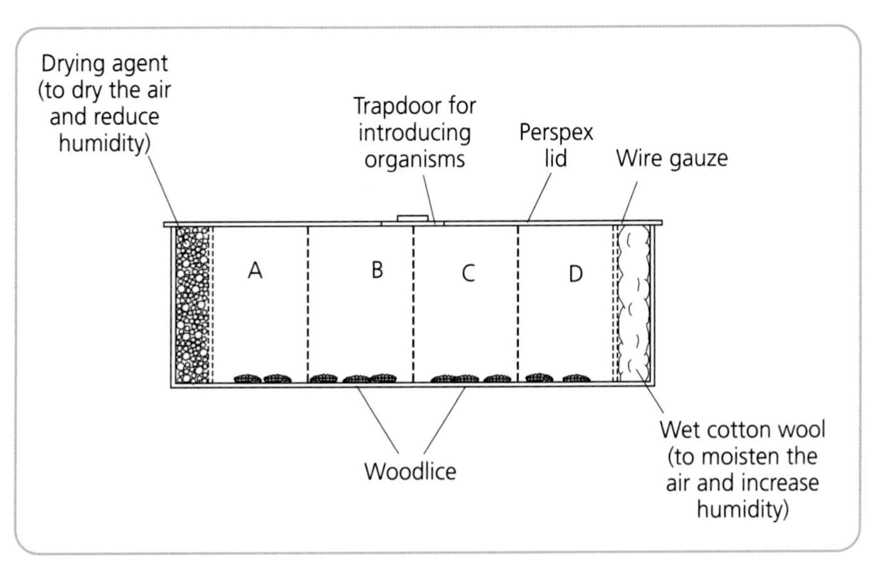

Figure 6.12

1 The response of woodlice in high humidity is to move slowly.

   The adaptive significance of this response is to increase the chance of remaining in a humid area and reduce the chance of drying out.

2 The response of woodlice in low humidity is to move quickly.

   The adaptive significance of this response is to increase the chance of reaching a humid area and reduce the chance of drying out.

   The apparatus in Figure 6.13 is used to demonstrate the behavioural response of woodlice to light and dark.

   Results from such investigations show the following.

   The response of woodlice is to move away from the light.

   The adaptive significance of this response is to increase the chance of reaching a dark area. Dark areas, in the normal environment of woodlice, are moist. Thus, this behavioural response reduces the chance of drying out.

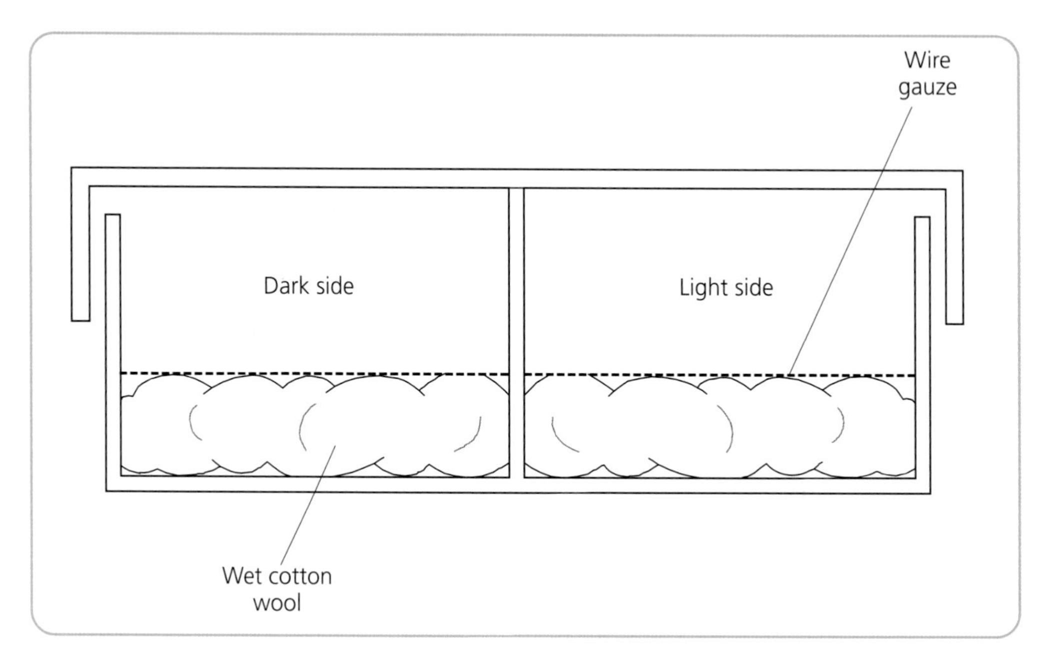

Figure 6.13

Stop Think Learn

1 Name two human activities that affect biodiversity and give an example of each.

2 Copy and complete this table on woodlice behaviour.

Environmental condition	Response of woodlice	Adaptive significance
Low humidity		
Light		

Hints and Tips

Make 'flash cards' for all the words and phrases in **bold** in the text.

In questions on consequence of removal of a species from a food web, remember to highlight the links between the organisms mentioned.

Questions

Examination Style Questions

1 Which of the following are adaptations of a desert plant?

A Small leaf surface area relative to volume; thick waxy cuticle; extensive shallow root system.

B Large leaf surface area relative to volume; thin waxy cuticle; extensive shallow root system.

C Large leaf surface area relative to volume; thin waxy cuticle; extensive shallow root system.

D Small leaf surface area relative to volume; thick waxy cuticle; short shallow root system.

2 Which of the following will maintain species diversity?

A very high intensity of grazing

B high intensity of grazing

C low intensity of grazing

D very low intensity of grazing

3 In an ecosystem, a community is

A all the animal species present.

B all the animal and plant species present.

C the total number of one species present.

D the total number of species present.

4 Figure 6.14 represents a pyramid of biomass.

Which letter represents the total biomass of the primary consumers?

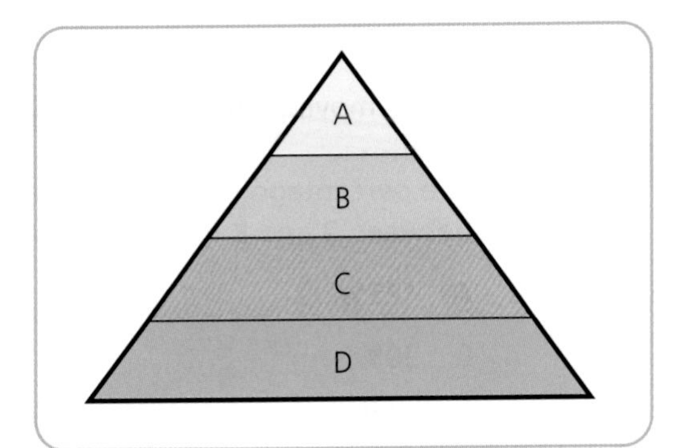

Figure 6.14

Questions 5 and 6 are based on an investigation into the response of woodlice to a difference in humidity (Figure 6.15).

Questions continued

Ten woodlice were placed into the choice chamber at point X. The lid was transparent to allow the woodlice to be seen.

The distribution of the woodlice were recorded every 2 minutes for a period of 10 minutes. The results are shown in Figure 6.16.

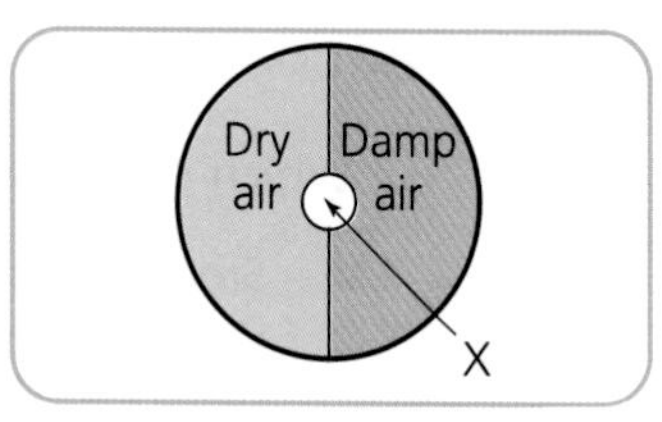

Figure 6.15

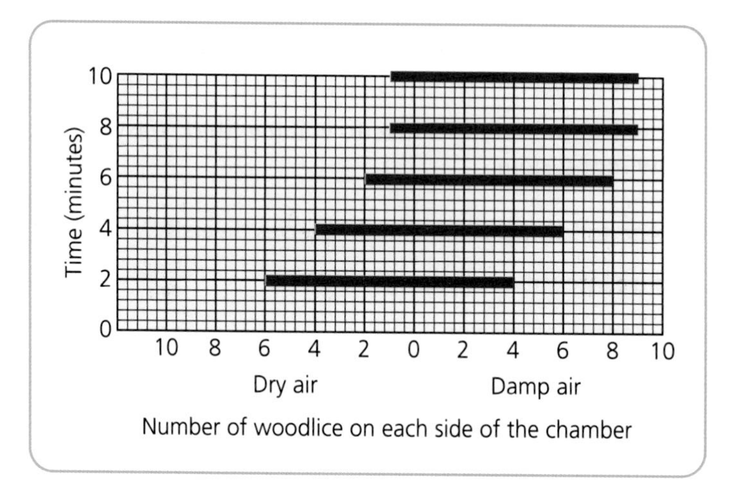

Figure 6.16

5 A conclusion from the investigation is that woodlice show a response to a difference in humidity

A by moving to a dry area.

B only in the dark.

C only in the light.

D by moving to a damp area.

6 The percentage increase in the number of woodlice in the damp side between 2 and 8 minutes was

A 125%.

B 90%.

C 50%.

D 5%.

7 Table 6.3 shows the changes in the number of plant species in a grassland ecosystem over 24 years with an increase in the intensity of grazing by sheep.

Questions continued

Years	Intensity of grazing	Average number of plant species present		
		Tree seedlings	Herbs	Grasses
1981–88	Low	2	8	10
1989–96	High	0	52	16
1997–04	Very high	0	16	4

a) Which two intensities of grazing by sheep decreased plant diversity? *(1)*

b) Name two human activities that lead to a decrease in species diversity. *(2)*

8 Figure 6.17 represents part of a food web in an ecosystem.

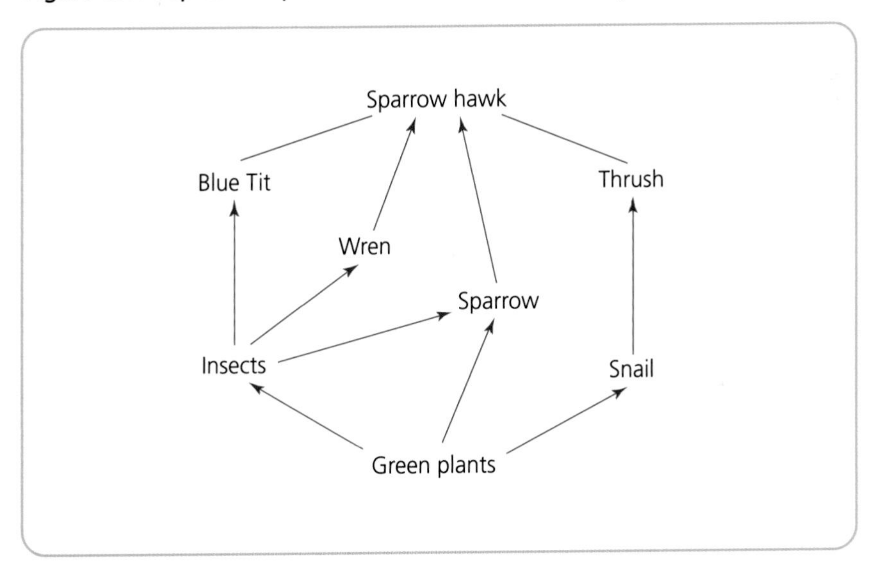

Figure 6.17

a) Use the food web to explain the meaning of community. *(1)*

b) Describe the niche of the snail in this food web. *(1)*

c) Identify an omnivore from the food web. *(1)*

d) Explain how the decline in sparrow hawk population leads to a decrease in the insect population. *(1)*

e) From the food web, explain why sparrows and wrens are in competition. *(1)*

9 Describe the behavioural adaptations shown by woodlice to humidity and light and the significance of these adaptations. *(5)*

Answers

Answers to examination style questions with commentary

1 You should know that desert plants have leaves with a small surface area, a thick waxy cuticle, extensive shallow roots and deep roots. **Answer = A**

2 You should know that very high and low intensities of grazing reduce diversity. And that high maintains diversity. **Answer = B**

3 You should know that community is all the species present. **Answer = B**

4 Base of pyramid are producers. C must be primary consumers. **Answer = C**

5 Results show move to damp air. Light/dark were not involved in the set up of the investigation. **Answer = D**

6 Time is shown by the y-axis. Read numbers from the x-axis on the damp side.

Number at 2 minutes = 4; Number at 8 minutes = 9; Increase = 5

Expressed as a % increase $\frac{\text{Increase}}{\text{Start}} \times 100\% = \frac{5}{4} \times 100\% = 125\%$ **Answer = A**

7 a) Add the numbers of each type at each intensity.

Low = 18; high = 68; very high = 20 **Answer = Low and very high**

b) Activities are pollution and habitat destruction. 1 mark each. **Do not be** tempted to answer by giving examples of pollution or habitat destruction.

8 a) You should know that community is **all the species of plants and animals in the ecosystem**.

b) Niche is shown by what it feeds on and what feed on it.

The niche of the snail is it feeds off green plants and it is the food for thrush.

c) Look for the organism that feeds off both plants and other animals = **Sparrow**.

d) Highlight all the links from sparrow hawk to insects. The birds are linked to the hawks and the insects are linked to the birds.

Answer = Fewer birds eaten by hawks and thus more birds to eat insects.

9 Describe the behavioural adaptations shown by woodlice to humidity and light and the significance of these adaptations.

In extended writing you **must** identify all the parts that make up the question.

Areas are:

- Describe behavioural adaptations to humidity.
- Describe the significance of these adaptations.

Answers continued

- Describe behavioural adaptations of woodlice to light.
- Describe the significance of these adaptations.

A1 Move slowly in humid/damp conditions.

B1 Increased chance of remaining in damp/humid conditions/area.

A2 Move rapidly in low humidity/dry conditions.

B2 Increased chance of reaching damp area/humid conditions.

B3 Damp/humid conditions reduce water loss/prevent drying out.

C1 Move away from light source.

B4 Increased chance of reaching a dark area.

B5 Dark area more likely to be damp/humid.

Maximum 2 marks from Group A.

Maximum 3 marks from Group B.

Maximum 1 mark from Group C.

NB Maximum score is 4 marks if point C1 is not answered.

Maximum Total 5 marks.

Chapter 7

FACTORS AFFECTING VARIATION IN A SPECIES

Continuous and Discontinuous Variation

Variation is differences within the same characteristic. Examples include flower colour, human height, eye colour, seed coat colour, blood group, seed shape, etc.

Variation is classified into two types.

Example

1 **Continuous variation** shows a continuous range of values for the same characteristic.

Characteristics include tree height, length of a limpet shell, shoe size, etc.

Continuous variation is represented by a histogram (see Figure 7.1).

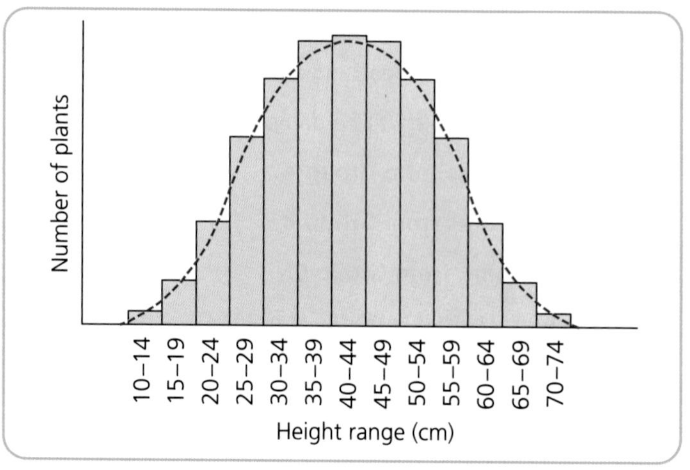

Figure 7.1 Continuous variation

2 **Discontinuous variation** shows two or more distinct forms for the same characteristic.

Characteristics include blood groups, flower colour, ability to roll tongue, etc.

Discontinuous variation is represented by a bar chart (see Figure 7.2).

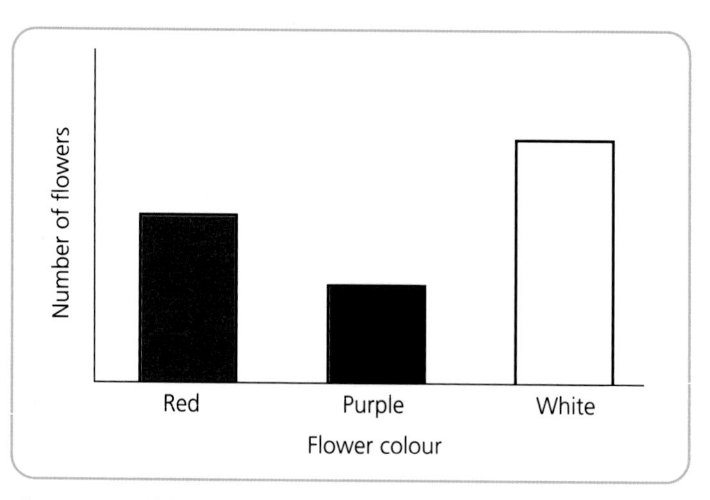

Figure 7.2 Discontinuous variation

Stop Think Learn

1 Figure 7.3 shows variation in body length in a sample of 250 young trout.

a) What type of variation is shown and justify your choice of answer?

b) How many fish in the sample had a body length of 6 cm or greater?

c) What was the range in body length in the sample?

d) Calculate the percentage of fish in the sample that had a body length in the range 4.0–4.4 cm.

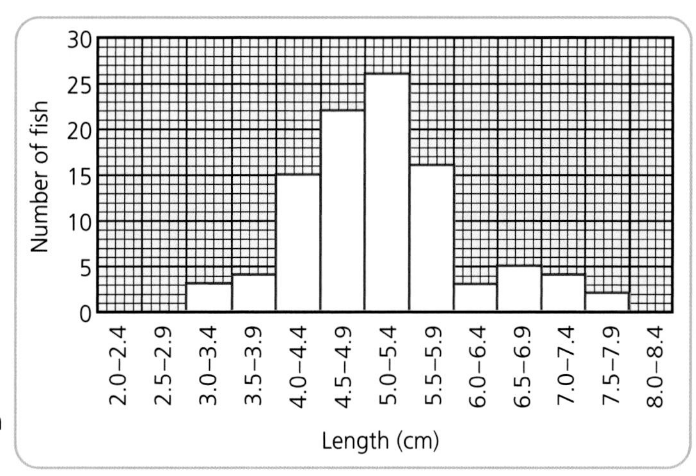

Figure 7.3

Fertilisation

Gamete production

Key Points

- Specialised cells are produced within the reproductive tissue to form the sex cells.
- Sex cells are called **gametes**. Gametes in animals are called **sperm** (male gametes) and **ova** (female gametes).
- Sperm are produced in the **testes**. Ova are produced in **ovaries** (*ovum singular*). Figure 7.4 shows the position of the testes.

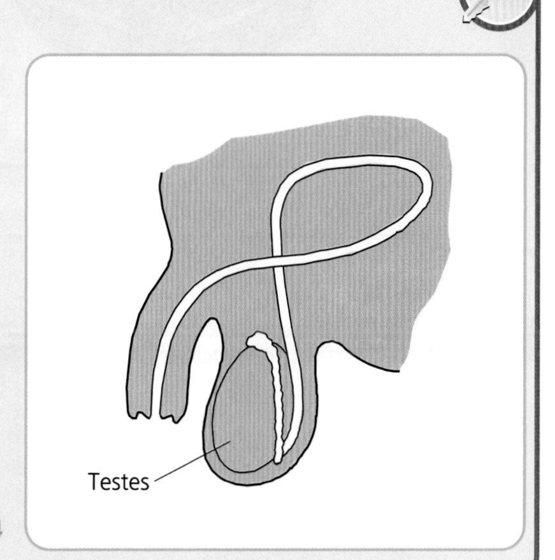

Figure 7.4

***Key Points** continued* ➢

Key Points continued

Figure 7.5 shows the position of the ovaries.

- In plants, a male gamete develops in a **pollen grain**.
- Pollen grains are produced in **anthers**.
- In plants, a female gamete develops in an **ovule**.
- Ovules are produced in the **ovary**.

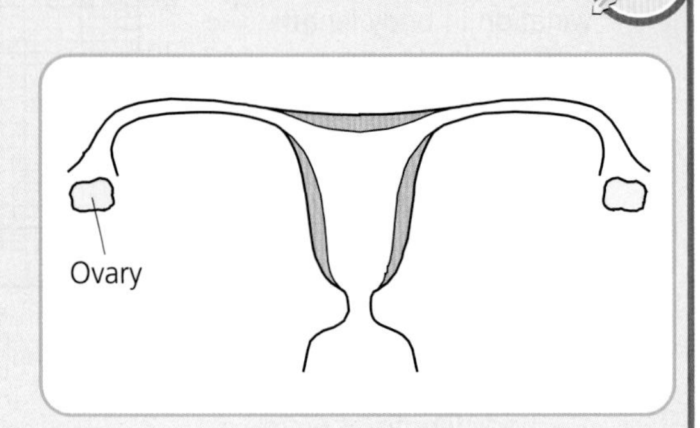

Figure 7.5

Figure 7.6 shows the position of the ovary, ovule and anthers in a flower.

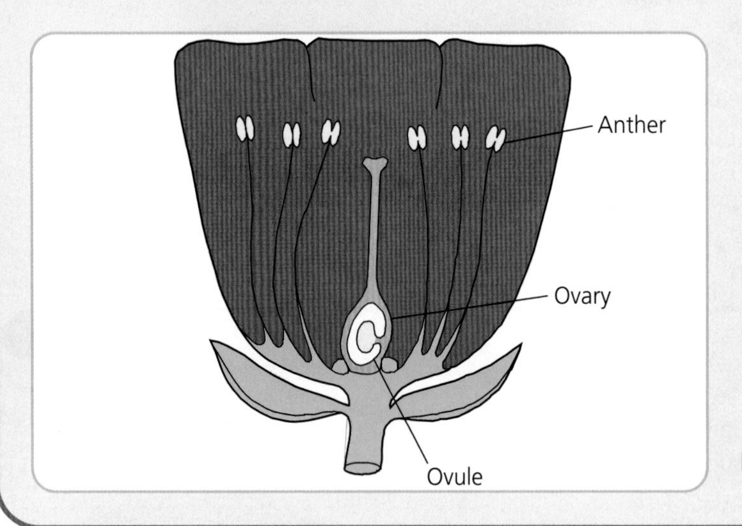

Figure 7.6

Fusion of nuclei

Key Points

- At **fertilisation**, the nucleus of a male gamete fuses with the nucleus of a female gamete.
- The single cell formed at fertilisation is the **zygote**.
- The zygote grows and develops into the young plant or animal.

Figure 7.7 shows formation of the zygote.

- The zygote shows variation as the gametes are produced by different parents.

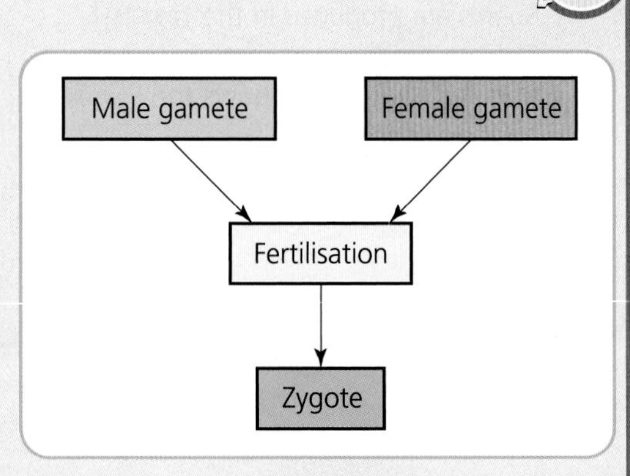

Figure 7.7

Stop Think Learn

1 Copy and complete this table to show gamete production in an animal.

Sex of animal	Name of gamete	Site of gamete production

2 Copy and complete the blanks in the sentences on gamete formation in plants.

A male gamete develops in a _______ _________ produced by an ________.

A ________ gamete develops in an _______ produced in the ovary.

3 Describe the process of fertilisation.

4 On a blank sheet make a flowchart to show how gametes form the zygote.

Division of the nucleus in gamete production – meiosis

Key Points

- Body cells have two matching sets of chromosomes.
- One set was donated in the gametes of each parent.

Figure 7.8 shows one pair of matching chromosomes of a fruit fly.

- A specific length of the chromosome is called a **gene** and each gene determines a characteristic of the organism.
- Genes for the same characteristic are on the exact same position of the chromosome, for example, antennae length.
- The form the gene expresses may differ, for example, black or red eye.

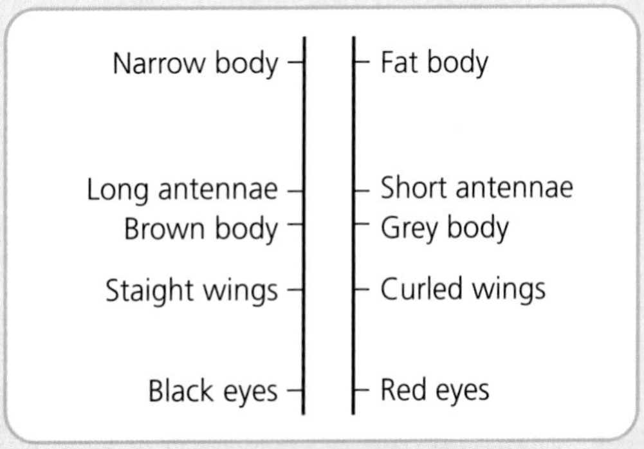

Narrow body – Fat body
Long antennae – Short antennae
Brown body – Grey body
Staight wings – Curled wings
Black eyes – Red eyes

Figure 7.8

- Specialised body cells present in the testes, ovaries and anthers are used to form gametes.
- During gamete formation, the chromosome number is reduced from two sets to a single set.
- **Meiosis** is the process that results in the reduction division.

Figure 7.9 (A and B) shows the two ways in which two pairs of matching chromosomes can align and separate during meiosis.

Shaded chromosomes originally came from one parent and unshaded from the other parent.

***Key Points** continued ➢*

Key Points continued

Figure 7.9A shows that the chromosomes from the original parents have aligned (lined up) on the same side *(Stage 1)* and that when the matching pairs of chromosomes separate from each other *(Stage 3)* the original parental chromosomes will be together in the same gamete.

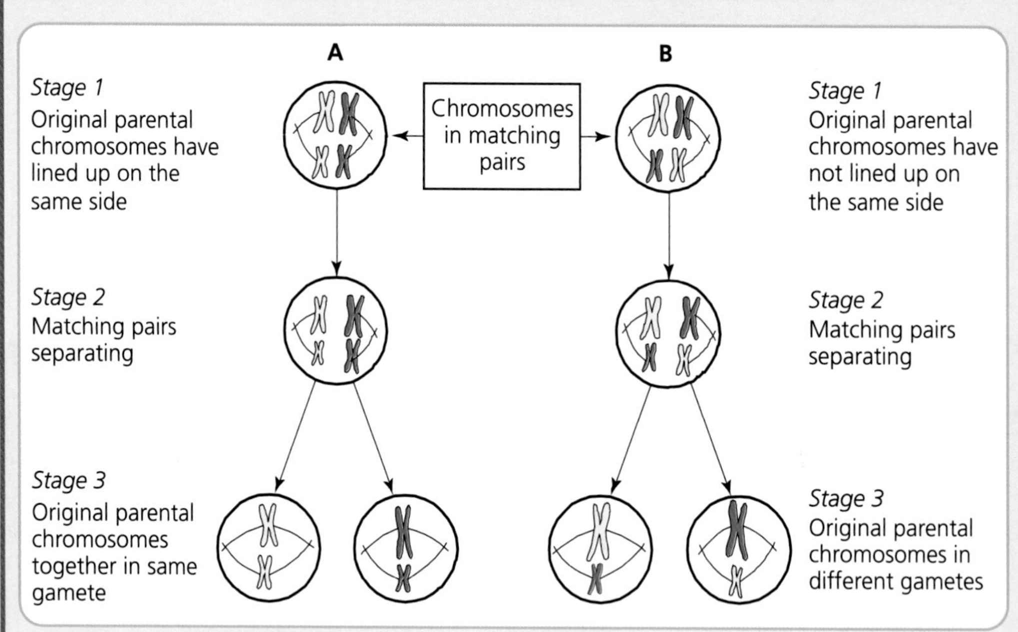

Figure 7.9

Figure 7.9B shows that chromosomes from the original parents have not lined up on the same side *(Stage 1)* and that when the matching pairs of chromosomes separate from each other *(Stage 3)* one of the chromosomes from each of the original parents will be in the same gamete.

- Matching pairs of chromosomes align at random in meiosis before separating.
- This is called **random assortment** of chromosomes.
- Random assortment gives rise to genetic variation in the gametes.
- If the gametes show variation, then, all the offspring will show variation.

Stop Think Learn

1. During meiosis, what happens to the chromosome number?
2. Show the four ways that three pairs of matching chromosomes can align in random assortment. Use shaded and unshaded chromosomes to represent the original parental chromosomes.
3. Name a source of variation in meiosis.

Chromosome structure

Key Points

- The genetic material on chromosomes determines the characteristics of an organism.
- Proteins determine the main characteristics of an organism
- The genetic material is a chemical called deoxyribonucleic acid (**DNA**).
- The genetic material exists as a chain of DNA bases.
- The order of DNA bases encodes the genetic information (Figure 7.10).
- The order of bases on DNA encodes information for the order of amino acids in a protein.
- The order of amino acids in a protein determine the structure of the protein.
- The structure of a protein determines its function.

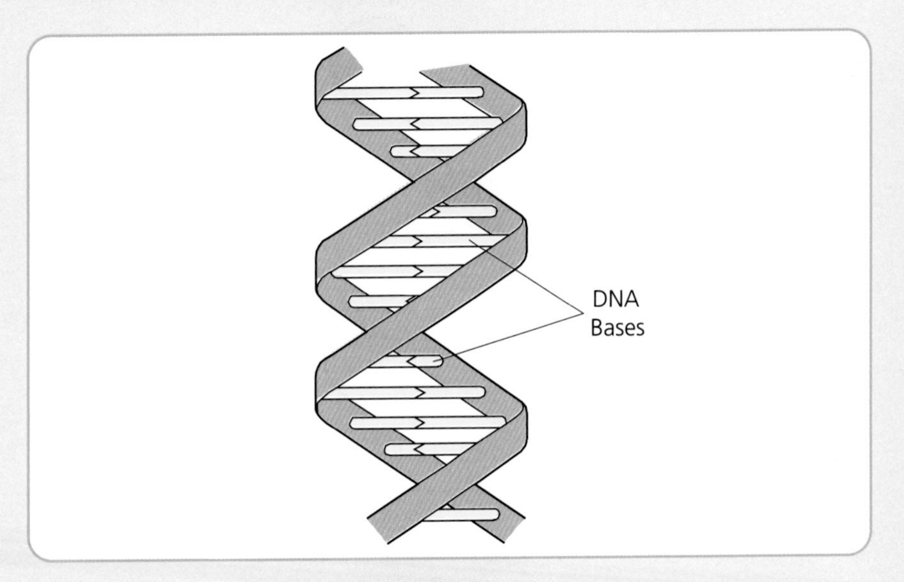

Figure 7.10

An example of this control is that genetic material, DNA, in a red blood cell controls production of the protein haemoglobin and the structure of haemoglobin allows it to combine with oxygen for transport.

- **The order of bases on DNA** determines the function of the protein.
- There are discrete lengths of chromosomal DNA that determine the production of different proteins. These discrete lengths of DNA are the **genes**.

Stop Think Learn

1 How is genetic information encoded in a chromosome?

2 Construct a flowchart to show the link between DNA, the structure of a protein and the function of a protein.

Chromosome numbers in different species

Key Points

- The one set chromosome number in humans is 23 chromosomes.
- Body cells have two sets and therefore 46 chromosomes are present.
- The number of chromosomes for a species is called the **chromosome complement**. **Human two set number** = 46 and **one set number** = 23.

Sex determination

Figure 7.11 outlines sex determination in humans.

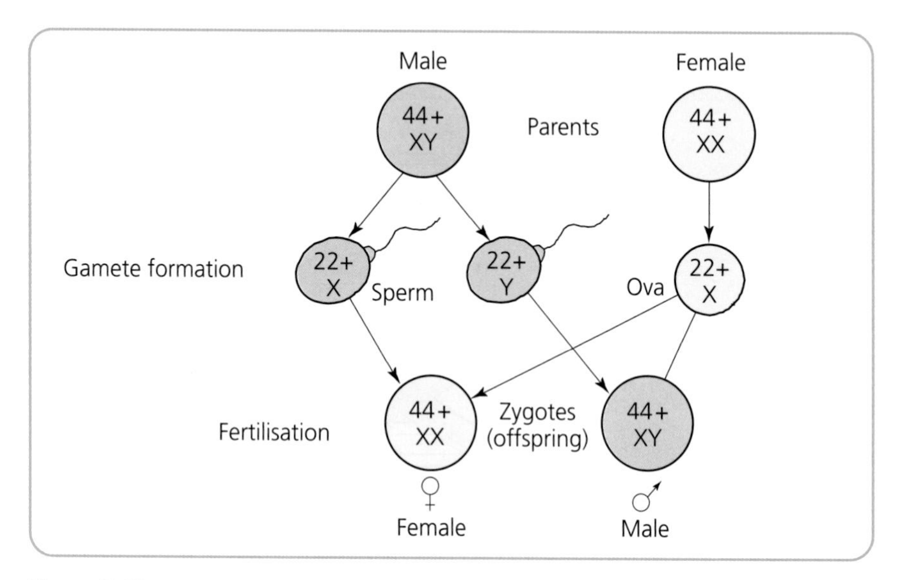

Figure 7.11

Key Points

- A pair of chromosomes determines sex in humans.
- Two X chromosomes determines a female.
- One X and one Y chromosome determines a male.
- In gamete formation, the sex chromosomes align and separate in the same way as other chromosomes. A female is XX and produces only **female gametes** that contain an X chromosome.
- A male is XY and produces **male gametes** that contain either an X or a Y chromosome.

Stop Think Learn

1. A cell of a cabbage leaf was found to contain 18 chromosomes. How many chromosomes are present in the nucleus of a pollen grain? Explain your answer.
2. With reference to the sex chromosomes show, by means of a diagram, the types of gametes that are formed in meiosis from a human female and a human male.

Hints and Tips

Visit the BBC Higher Biology Bitesize website for animation on meiosis (www.bbc.co.uk/scotland/education.bitesize/).

To identify a type of variation, ask yourself if the characteristic had continued to change throughout the lifetime of the organism. If the answer is YES it is continuous variation if NO then it is discontinuous variation.

To remember sets of chromosomes – you got one set from each parent so your body cells must have two sets.

Remember female humans have 23 pairs of matching chromosomes whereas males have only 22 pairs of matching chromosomes and a pair of sex chromosomes, X and Y that do not form a matching pair.

Enzymes are proteins that control chemical reactions.

Many hormones are proteins. Hormones regulate body functions and control growth and development.

Make 'flash cards' for all the words and phrases in **bold** in the text.

Questions

Examination style questions

1. The list is of some human variations.
   1. size of feet
   2. adult height
   3. free or attached ear lobes
   4. ability to roll tongue

   Which of the above show continuous variation?

   A 1 and 2 only

   B 1 and 3 only

   C 2 and 4 only

   D 3 and 4 only

Questions continued

2 In which of the following animal structures will meiosis take place?

A skin and ovary

B testes and skin

C testes and ovary

D liver and skin

3 The sites of production of male and female gametes in a flowering plant are:

A anther and ovary.

B anther and pollen.

C ovule and pollen.

D ovary and ovule.

4 The following statements refer to human chromosome numbers.

1 The one set chromosome number is 23.

2 The nucleus of a normal body cell contains 46 chromosomes.

3 The two set chromosome number is restored during meiosis.

Which statements are correct?

A 1 and 2 only

B 1 and 3 only

C 2 and 3 only

D 1, 2 and 3

5 Copy and complete the table to show the structures that contain gametes and their site of production in flowering plants.

Type of gamete	Structure containing gamete	Site of production
Female		ovary
Male	pollen grain	

6 Underline one option in each set of brackets to make the sentence correct.

It is the (number / order) of DNA bases that encodes information for the sequence of amino acids in (chromosomes / proteins).

7 Copy the table at the top of page 81. Decide whether each of the following statements is TRUE or FALSE and tick the appropriate box.

If the statement is FALSE, write the correct word in the Correction box to replace the word underlined in the statement.

Questions continued

Statements	True	False	Correction
The <u>number</u> of DNA bases encodes information for the sequence of amino acids in a protein.			
Chromosomes are made up from chains of <u>DNA</u> bases.			
In humans <u>female</u> gametes contain either an X or a Y chromosome.			

Answers

Answers for examination style questions with commentary

1 Continuous variation shows a wide range – size of feet; adult height **Answer = A**

2 You should know meiosis takes place in testes and ovaries in animals. **Answer = C**

3 You should know that in plants gametes are produced in the ovule and pollen. **Answer = C**

4 You should know that only statements 1 and 2 are correct. The two set number is restored at fertilisation not meiosis. **Answer = A**

5 You should know that a gamete develops in an **ovule** and that pollen is produced in the **anther**. *(2 × 1 mark)*

6 You should know that it is the **<u>order</u>** in **<u>proteins</u>**. *(1 mark)*

7 You should know that it is the order and not the number of bases. **False order**

You should know that this is correct. **True**

You should know that male gametes have either an X or Y. **False male** *(3 × 1 mark)*

Chapter 8

GENETICS

Terms Used in Genetics

Key Points

- **Genetics** is the study of the inheritance of genes from parents to offspring.
- A **gene** is a specific length of chromosomal DNA that determines a characteristic of the organism.

Figure 7.8 (p. 75) shows one pair of matching chromosomes from the two sets of matching chromosomes of a fruit fly.

- Genes for the same characteristic are in the exact same position on the chromosome, for example, antennae length.
- The form that a gene expresses can differ, for example, black or red eye, straight or curled wings.
- The different forms of a gene are called **alleles**.
- Body cells have two copies of each gene as they have two sets of matching chromosomes.
- Gametes have one copy of each gene as they have only one set of chromosomes.
  Here is a list of terms used in genetic crosses and their definition.

  Genotype: The genetic make up of an individual.

  Phenotype: The appearance of an individual.

  Dominant: In individuals with two different alleles of a gene, the form that is expressed as the phenotype is dominant.

  Recessive: In individuals with two different alleles for a gene, the form that is **not** expressed as the phenotype is recessive.

  Homozygous: An individual having an identical pair of alleles of a gene.

  Heterozygous: An individual having two different alleles of a gene.

  True-breeding: Individual is homozygous and both alleles are for the same genotype.

  Cross: Interbreeding of two individuals to produce offspring.

  P: Identifies the parents in a cross.

  F_1: Identifies the offspring in the first generation in a cross.

  F_2: Identifies the offspring in the second generation in a cross.

Key Points continued ➢

Key Points continued

Monohybrid cross: A cross in which the inheritance of one pair of alternative characteristics is followed, for example, long and short antennae.

Nomenclature: Letters are used to identify the alleles of genes. The dominant allele is identified using the upper case of the letter that starts it and the recessive allele is identified using the lower case of the letter that starts the dominant allele, for example, long dominant to short, then L for long and l for short.

Example of a Monohybrid Cross to the F_2 Generation

Hints and Tips

As you work your way through the cross, check on the list of definitions to reinforce your understanding of the cross.

For example, when you are told that an individual is true-breeding this means that they are homozygous and that both alleles determine the same phenotype.

Remember that each individual has two alleles of each gene in their body cells.

Remember that in gamete formation the matching pairs of chromosomes are separated, thus there is only one allele of each gene in gametes.

Example

In pea plants, the gene for height has two alleles. The allele for tall is dominant to the allele for dwarf. True-breeding tall plants were crossed with true-breeding dwarf plants. The F_1 plants were then crossed to produce the F_2 generation.

P	Phenotype	Tall	X	Dwarf
P	Genotype	TT	X	tt
P	Gametes	T		t
F_1	Genotype		Tt	
F_1	Phenotype		All Tall	

Plants of the F_1 are then crossed

P	Phenotype	Tall	X	Tall
P	Genotype	Tt	X	Tt
P	Gametes	T or t		T or t

Example continued ➢

Example continued

Gametes T and t are produced in equal numbers as a result of separation of matching pairs of chromosomes in meiosis.

The gametes combine at random and to ensure that you get the combinations correct, you use a Punnett square (Figure 8.1).

The two possible gametes of each parent is shown in the shaded areas; the possible combinations of these gametes is shown in the unshaded areas

Gametes	T	t
T	TT	Tt
t	Tt	tt

Figure 8.1 Punnett square

From the results:

Expected F_2 phenotype ratio is 3 Tall : 1 Dwarf.

Expected F_2 genotype ratio is 1 TT : 2 Tt: 1 tt.

In crosses, the expected ratios may not be obtained for the following reasons:

1 Sample size is too small.
2 Fertilisation is a chance / random process.

Stop Think Learn

Outline the following monohybrid crosses in the same way as the example.

1 In a species of plant the gene for flower colour has two alleles. The allele for red flower is dominant to the allele for white flower. Show the expected phenotype and genotype ratios that would be obtained from the cross of a true-breeding red flowering plant with a true-breeding white flowering plant through to the second generation. The second generation was obtained from crosses between plants of the first generation.

2 Black coat in mice is dominant to brown coat. Show the cross to the first generation of a true-breeding black mouse with a brown mouse and then interbreeding individuals from the first generation to produce a second generation.

Co-dominance

Figure 8.2 shows that in Shorthorn cattle, if a red cow is mated with a white bull the calves have a roan coloured coat.

When the coats of the calves are examined they are seen to have both red and white coloured hairs.

Neither red nor white is recessive.

In examples of this type in which both alleles are expressed, the alleles are described as being **co-dominant**.

The nomenclature for co-dominance is to use the upper case letter for both alleles.

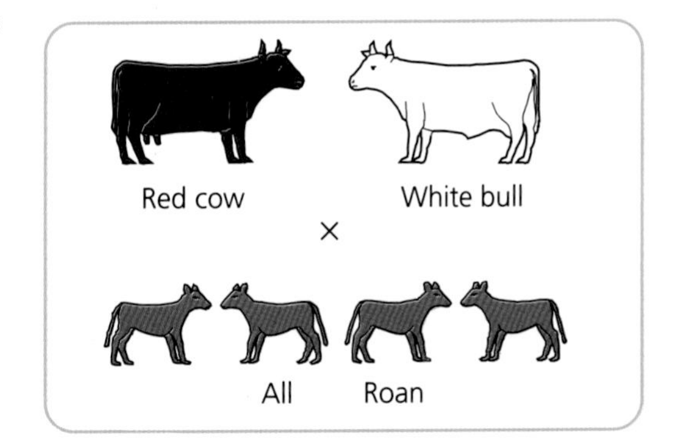

Figure 8.2

Examples

	Example 1			Example 2			
P Phenotype	Red cow	X	White bull	Roan cow		X	Roan bull
P Genotype	RR	X	WW	RW		X	RW
P Gametes	R		W	R or W			R or W
F_1 Genotypes		RW		RR	RW	RW	WW
F_1 Phenotypes		All Roan		1 Red		2 Roan	1 White

Hints and Tips

These are expected results. Results can vary from the expected due to the fact that fertilisation is a chance process or that the sample size is too small.

Polygenic Inheritance

Some characteristics are controlled by the interaction of the alleles of several genes.

The interaction of the alleles results in a range of phenotypes for that characteristic.

Examples of polygenic inheritance include skin colour in humans and seed mass in plants.

Polygenic inheritance shows continuous variation.

Environmental Impact on Phenotype

The final appearance of an organism is the result of the interaction of the genotype and the effects of the environment. This is represented by the equation:

Phenotype = Genotype + Environment

For example, two human individuals with the exact same genotype for skin colour go on holiday. One goes to sunny Spain for a beach holiday and the other one stays in Scotland. Due to differences in the environment, one has a golden tan while the other remains with an unchanged phenotype.

Hints and Tips

Individuals with the appearance of the dominant allele could be either homozygous dominant or heterozygous.

An allele can be shown to be dominant only by observing the phenotype of a heterozygous individual. Both forms of the allele are in this individual.

The only individual that you can tell their genotype by observing their phenotype is one with the appearance of the recessive allele. If red is dominant to white then a white individual must be **rr**.

In some problems to find the genotypes of the parents, the clue is in the fact that one of the offspring is homozygous recessive. This means that they must have received one copy of the recessive allele from each parent.

The prefix **poly** means many. Polygenic = many genes involved.

Make 'flash cards' of the words and phrases in **bold** in the text.

Questions

Examination style questions

1 Allele is a term used to describe:

A the different phenotypes of a species.

B the possible gametes produced by an individual.

C the different genotypes in a population.

D the different forms of a gene.

2 The term used to describe the inheritance of a characteristic that is controlled by the alleles of several genes is:

A co-dominant.

B monohybrid.

C polygenic.

D natural selection.

Questions continued

3 In a species of fly, grey body is dominant to black body. A true-breeding grey fly was crossed with a black fly. Female offspring of the F_1 generation were interbred with black-bodied males. Which of the following is the expected phenotype ratio in the F_2 generation?

A all black

B all grey

C 1 grey : 1 black

D 3 grey : 1 black

4 True-breeding tall pea plants were crossed with true-breeding dwarf plants.

All the F_1 plants were tall.

a) Using appropriate symbols for the alleles, copy and complete the following to show the genotypes of the parents and offspring:

P phenotypes	Tall	X	Dwarf
P genotypes	_____		_____
F_1 phenotype		Tall	
F_1 genotype		_________	*(1)*

b) Plants of the F_1 generation were allowed to self-cross. What would be the expected ratio of phenotypes in the second generation? *(1)*

c) When the plants were counted there were 360 tall plants and 90 dwarf plants.

Calculate the ratio of tall to dwarf plants obtained in the cross. *(1)*

d) Explain why the results differ from the expected results.

5 In shorthorn cattle, coat colour is controlled by a single gene. The allele for red coat (R) is co-dominant with the allele for white coat (W). The heterozygous condition results in a coat consisting of a mixture of red and white hairs and is described as being roan colour.

A bull of genotype RW was crossed with a cow of genotype WW.

a) State the phenotype of each animal. *(1)*

b) Copy and complete the following table by

i) inserting the genotype of the male and female gametes. *(1)*

ii) showing the possible genotype of the offspring. *(1)*

Questions continued

Female

Gametes		

Male

c) Show the expected phenotype ratio of the offspring. *(1)*

6 The fur on the face, ears, paws and tip of tail of Siamese cats living in a hot climate are light coloured. In the colder climate of Scotland, these areas of fur are dark coloured.

Use the information on Siamese cats to explain the meaning of the equation:

Genotype + Environment = Phenotype. *(2)*

Answers

Answers to Examination Style Questions with commentary

1 You should know that allele is the form of a gene. **Answer = D**

2 If a characteristic is controlled by several genes it is polygenic. **Answer = C**

3

1st cross	P phenotypes	Grey	X	Black
	P genotypes	GG	X	gg
	P gametes	G		g
	F_1 genotype		Gg	
	F_1 phenotype		All grey	
2nd cross	P phenotypes	Grey female	X	Black male
	P genotypes	Gg	X	gg
	P gametes	G or g		g
	F_2 genotype	Gg	or	gg
	F_2 phenotype	grey	black	

Expected ratio 1:1 **Answer = C**

4 a) Upper case of first letter for dominant = T and lower case for recessive = t.

Parents are true breeding, therefore homozygous. Tall = TT Dwarf = tt

Gametes would be T only from tall parent and t only from dwarf. F_1 = Tt

Answers continued

b) P phenotype Tall X Tall

P genotype Tt X Tt

P gametes T or t T or t

Gametes	T	t
T	TT	Tt
t	Tt	tt

Answer = Expected ratio 3 Tall : 1 Dwarf

c) Divide 360 and 90. **Answer = Ratio 4 Tall : 1 Dwarf**

d) You should know that this is because fertilisation is a chance process or that the sample size is too small.

5 a) Both alleles expressed gives roan. Bull = RW = Roan Cow = WW = White

b)

Gametes	W	W
R	RW	RW
W	WW	WW

Answer =
Gametes 1 mark
Genotypes 1 mark

c) 2 roan and 2 white offspring.

Ratio of 2 Roan : 2 White. Always simplify a ratio. **Answer = 1 Roan : 1 White**

6 Two marks, therefore two points required. The question asks to explain how genotype and environment interact. The environmental part has to be temperature (warning: do not use heat as an equivalent for temperature).

Answer = Genes control the colour of the fur *(1st mark)*

but temperature has an effect on the depth of the final colour. *(2nd mark)*

Chapter 9

SELECTION

Natural Selection

Key Points

- Natural selection refers to the theory of Charles Darwin. Figure 9.1 follows natural selection as outlined by Darwin and Figure 9.2 outlines natural selection as illustrated in a species of moth called the Peppered moth.

1. Population of a species shows variation
   ↓
2. The environment changes
   ↓
3. Best adapted to the environment are selected
   ↓
4. Best adapted survive
   ↓
5. Best adapted breed
   ↓
6. Favourable adaptation passed on and their numbers increase in the population

Figure 9.1

1. Variation in Peppered moth is shown by the light and dark forms
   ↓
2. Air pollution kills the lichens and darkens the bark with soot
   ↓
3. The dark form are better camouflaged and fewer are eaten by birds than the light form
   ↓
4. More of the dark form survive
   ↓
5. More of the dark form breed
   ↓
6. Number of individuals of the dark form increases in the population

Figure 9.2

- Peppered moths show two main variations: a light and a dark form.
- Moths are active at night and during the day they settle onto the bark of trees (Figure 9.3).
- Birds hunt for the moths during the day.

Key Points continued ➢

Key Points continued

- In unpolluted woodlands, the bark is light coloured as a result of the presence of lichens.
- In areas where the air is polluted, the lichens die and the bark is darkened with soot.

Figure 9.3

Stop Think Learn

1. Explain why it is important that a species shows variation.
2. Explain what is meant by 'best adapted to the environment are selected'.

Selective Breeding

Key Points

- Man carries out selective breeding.
- Man selects members of plant and animal species with the desired characteristics.
- Only individuals with the desired characteristics are allowed to interbreed.
- Selection is repeated over many generations.
- Selective breeding takes a long time (many generations).
- Selective breeding is not always guaranteed.
- Undesired characteristics can also be enhanced.
- Selective breeding is summed up as 'Select the best and hope for the best'.
- Examples of selective breeding include cattle for milk yield or meat yield, variety of breeds of dog (Figure 9.4), crops for their yield of fruit or seeds, colour, flavour, storage properties etc.

Figure 9.4

Genetic Engineering

Key Points

- **Genetic engineering** is when man identifies and removes a desired gene from one species and inserts the gene into another species. The 'engineered organism' synthesises the product of the inserted gene.
- Bacteria are widely used in genetic engineering.
- **Plasmids** are rings of DNA present in bacterial cells.
- Plasmids can be removed from bacterial cells, modified and then reinserted into bacterial cells.

Some stages in the production of a desired product by genetic engineering are outlined in Figure 9.5. The product is human growth hormone (GH).

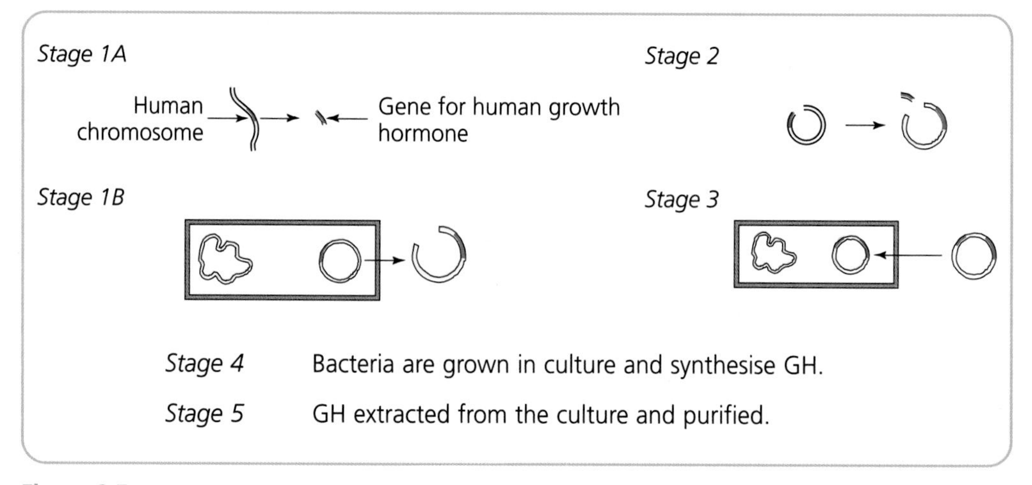

Figure 9.5

Stage 1A	Gene for GH is identified and removed from chromosome.
Stage 1B	Plasmid removed from bacterial cell and cut open.
Stage 2	Gene for GH is inserted into plasmids.
Stage 3	Modified plasmids inserted into bacterial cells.

Applications of genetic engineering include: production of medicines for human use, for example, GH and insulin.

GH is given to children under the expected range of heights for their stage of development.

Insulin is a hormone that reduces blood sugar levels. Individuals who do not produce sufficient insulin suffer from a form of diabetes.

Advantages of genetic engineering:

a) provides an increased range of products,

b) provides an increased rate of production.

Disadvantages of genetic engineering:

a) High cost of development of genetic engineering.

b) The possible release of genetically engineered bacteria into the environment that may not respond to available medicines.

Stop Think Learn

Use the information in the text to construct a flowchart to show genetic engineering.

The outline at the start is shown.

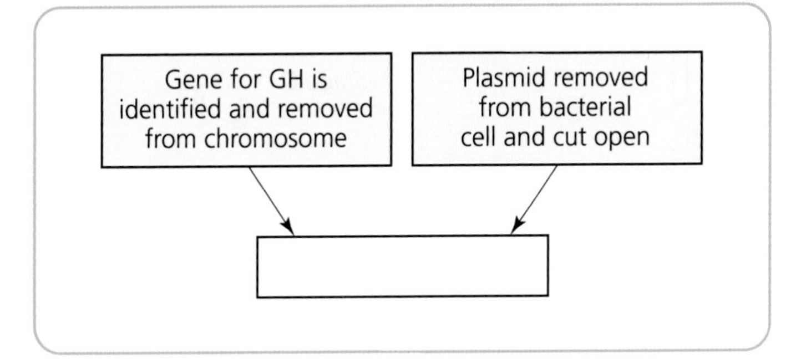

Hints and Tips

Natural selection requires a selecting agent. In the Peppered moth, it is birds feeding on moths that is the selecting agent.

Insulin is a protein and a hormone. Insulin reduces the blood glucose level by increasing glucose uptake from the blood by liver cells.

In genetic engineering, the removal of required genes and the cutting open and closing of plasmids must be carried out by enzymes.

Make 'flash cards' of the words and phrase in **bold** in the text.

Questions

Examination Style Questions

1 Figure 9.6 shows stages of the process of genetic engineering.

a) Describe what happens during stages B, D and E. *(3)*

b) Name a product that can be manufactured using genetically engineered bacteria. *(1)*

c) What type of chemical substance would be used to cut open a plasmid? *(1)*

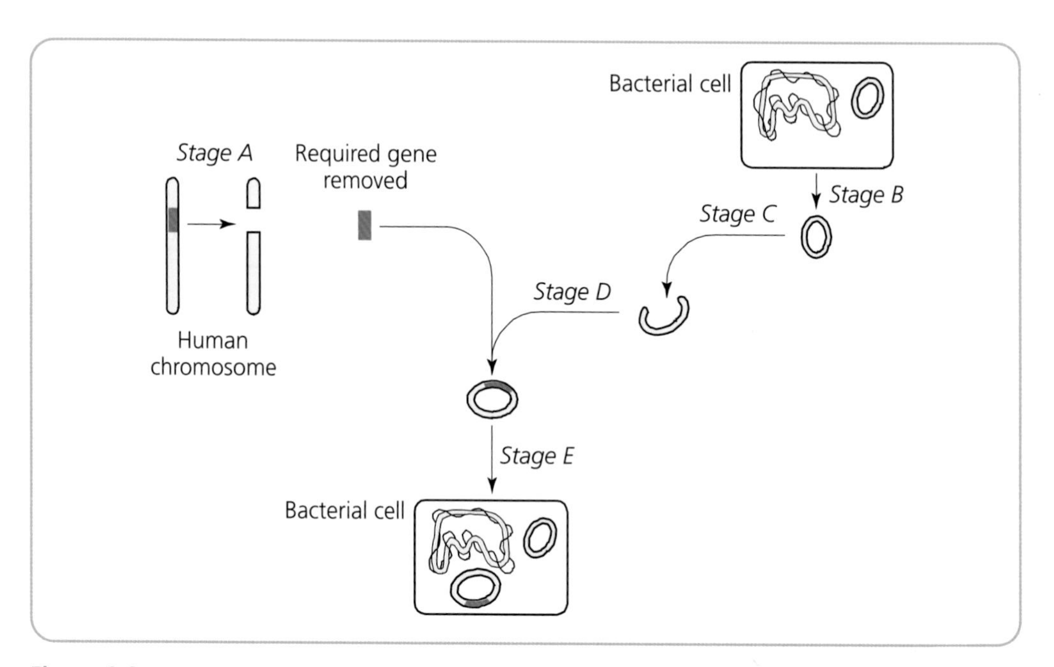

Figure 9.6

2 Using the Peppered moth to illustrate your answer, describe natural selection. *(5)*

3 Describe selective breeding. *(5)*

Answers

Answers to examination style questions with commentary

1 a) During stage B, plasmids are removed from bacterial cells.

During stage D, the required gene is inserted into the plasmid.

During stage E, plasmids with the required gene are inserted into bacterial cells.

b) You should know that products include human growth hormone and insulin.

c) You must realise this is a chemical change and these are carried out by enzymes.

2 Using the Peppered moth to illustrate your answer, describe natural selection.

In extended writing questions, you **must** identify all the parts that make up the question. These are:

- variation,
- environmental change,
- selection pressure,
- result of selection.

A1 Moth has light and dark forms.

A2 Bark of tree darkened due to air pollution.

A3 Moths rest on bark during the day.

Maximum of 2 marks.

B1 Dark form better camouflaged.

B2 Fewer of the dark form eaten by birds.

Maximum of 1 mark.

C1 More of the dark form survive.

C2 More dark moths breed.

C3 Increase in number with the dark form in the population.

Maximum of 2 marks. **Maximum Total = 5 marks.**

3 Describe selective breeding

In extended writing questions, you **must** identify all the parts that make up the question. These are:

- how selection takes place with advantages,
- possible disadvantages.

Examples

A1 Man selects individuals with desired characteristics.

A2 Only those with desired characteristics are allowed to breed.

Answers continued

A3 Selection repeated over several generations.

A4 Selective breeding takes a long time.

Maximum of 3 marks.

B1 Improvement not guaranteed.

B2 Undesired characteristics can also be enhanced.

Maximum of 1 mark.

C1 Named example of selective breeding.

Maximum of 1 mark. **Maximum Total = 5 marks.**

UNIT 3

Animal Physiology

Chapter 10

MAMMALIAN NUTRITION

Food Groups

Requirement for food

Key Points

- The **main food groups** are carbohydrates, proteins, fats, vitamins and minerals.
- The **role of carbohydrate** is to provide an energy source in respiration.
- The **role of protein** is to provide a source of amino acids for production of all the proteins of the body. These include enzymes, hormones, haemoglobin and antibodies.
- The **role of fat** is to provide an energy source in respiration and to provide an energy store.
- The **role of vitamins** is to give protection against disease and for regulation of metabolism. Vitamins are required for good health.
- The **role of minerals** include supplying iron for the formation of haemoglobin and calcium phosphate for the formation of bones and teeth.

The table below shows the structure of carbohydrates, proteins and fats.

Food group	Chemical elements present	Basic units
Carbohydrates	Carbon hydrogen and oxygen	Sugars, for example, glucose
Protein	Carbon hydrogen, oxygen and nitrogen	20 different types of amino acids
Fats	Carbon hydrogen and oxygen	3 fatty acids and glycerol

Figure 10.1 Protein, fats and carbohydrates

Figure 10.2 Glucose molecule

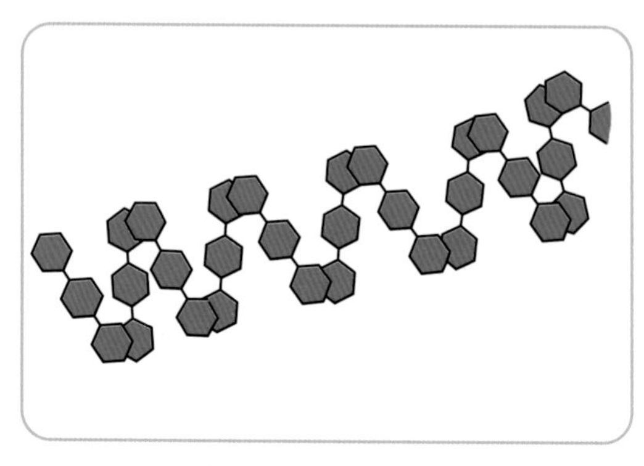

Figure 10.3 Glucose molecules bond to form chains of either starch, cellulose or glycogen

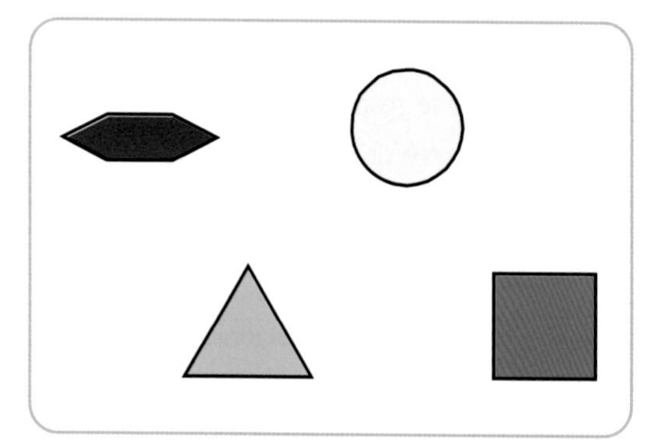

Figure 10.4 Amino acids

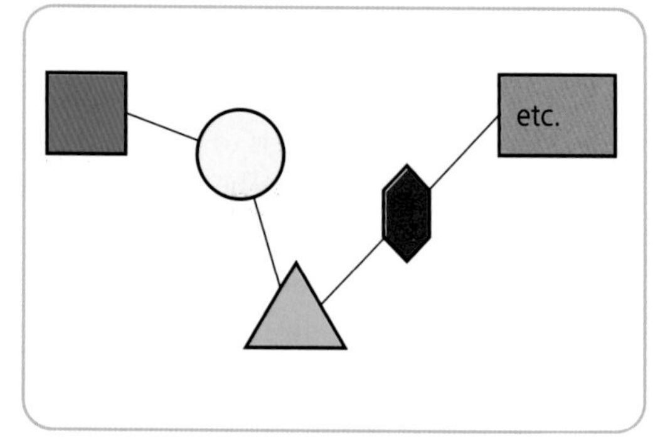

Figure 10.5 Each protein differs due to:
1 the amino acid sequence in the chain,
2 the length of the chain.

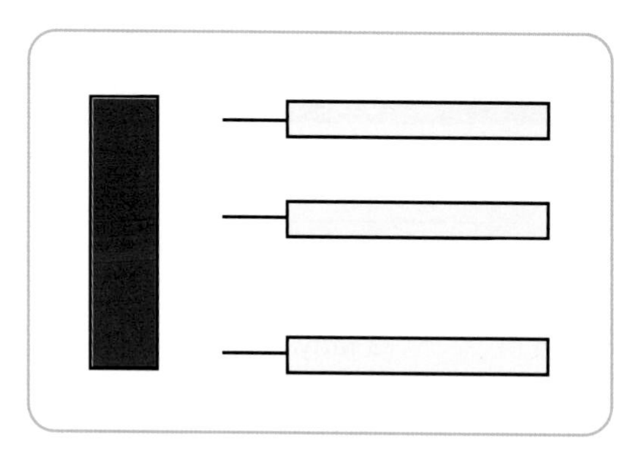

Figure 10.6 Glycerol and fatty acids

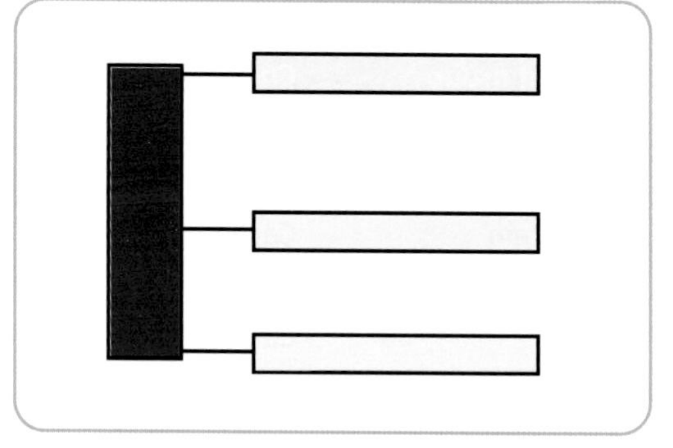

Figure 10.7 Fatty acids bond with glycerol to form fat

Stop Think Learn

Copy and complete the table to show the role of each of the five food groups.

Carbohydrate has been completed for you.

Food group	Role of food group in the body
Carbohydrates	Provide an energy source in respiration

Food Tests and Energy Content of Food

The table shows the tests for some of the food groups.

Food	Test	Positive result
Starch	Add iodine to the sample	Colour changes from colour of iodine to blue-black
Glucose	Heat sample with Benedict's reagent	Colour changes from blue to red
Protein	Add Biuret reagent to sample	Colour changes from light blue to violet
Fat	Rub sample against brown paper	Translucent spot develops on the paper

This table shows the energy content of the three main food groups.

Food group	Energy / g (kJ)
Carbohydrate	20
Protein	20
Fat	40

Weight for weight **energy in fats** is twice the energy of carbohydrates and protein.

The Need for Digestion

Figure 10.8 represents digestion in the alimentary canal.

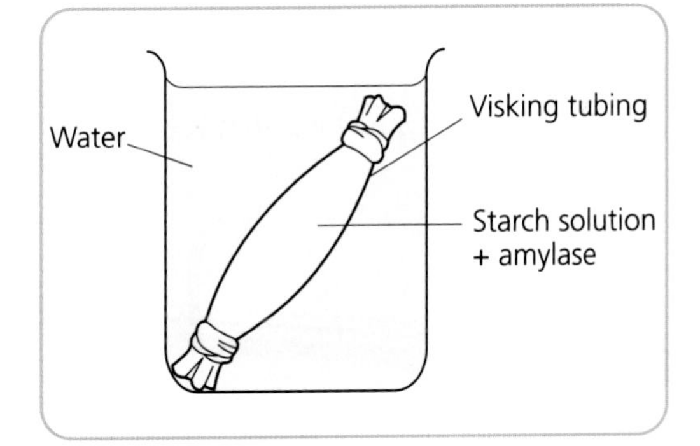

Figure 10.8

Visking tubing has small pores that allow small sugar molecules to pass through but prevent the passage of large molecules of starch.

In the model, the visking tubing represents the walls of the small intestine.

The starch solution and amylase represent food and digestive enzymes.

The water represents the surrounding blood capillaries.

Results of testing water for sugar and starch

Starch never appears in the water.

After five minutes sugars are present in the water.

Digestion is needed because large insoluble molecules in the food must be broken down into small soluble molecules by the action of enzymes. This digestion allows the small molecules to be absorbed through the wall of the small intestine into the blood stream.

Hints and Tips

Link protein structure and function to DNA (Chapter 7).

Link protein structure to the role of enzymes and hormones (Chapter 3).

Stop Think Learn

1. A food sample gave a blue colour with Biuret reagent, a red colour with Benedict's reagent, no clear spot on brown paper and a blue-black colour with iodine.

   Name the food classes present in the sample.

2. One hundred grammes of a food gave the following analysis.

   Water = 40 g, starch = 20 g, protein = 30 g, fats 10 g.

   Using values from the table on page 101, calculate the energy in 100 g of the food.

3. Copy and complete the sentences on the need for digestion.

   Large ___________ molecules cannot pass through the walls of the _____________.

   These molecules are converted to small _______ molecules by the action of _______.

   Thus digestion allows the molecules to be __________ into the ____________.

Structure and Function of the Alimentary Canal and Associated Organs

Refer to Figure 10.9 throughout this part of the work.

The KEY names structures that you need to identify in the alimentary canal.

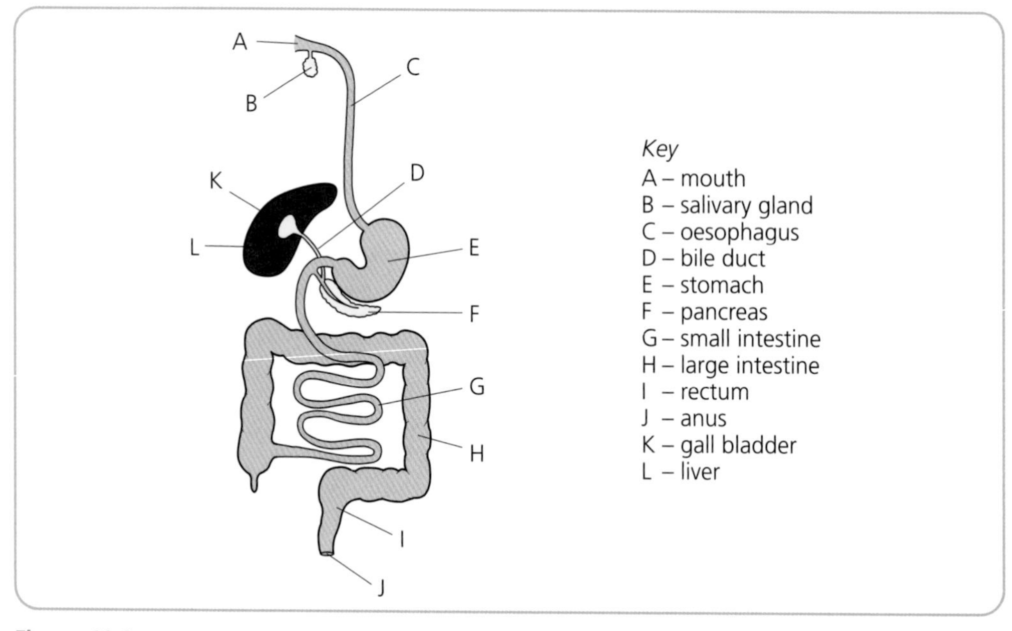

Figure 10.9

The role of the mouth, salivary glands and oesophagus in digestion

Key Points

- **Mechanical breakdown** of the food is by the action of the teeth.
- **Salivary glands** secrete saliva which is mixed with the food as it is chewed.
- **Mucus** in the saliva lubricates the food for swallowing and also the surface of the mouth to reduce friction.
- When the food is swallowed it is passed down the oesophagus to the stomach by **peristalsis**.
- **Amylase** in the saliva starts the breakdown of starch to maltose.

Figure 10.10 shows the mechanism of peristalsis.

- Muscles in the wall of the oesophagus immediately behind the food contract.
- At the same time muscles immediately in front of the food relax.
- Contraction of the muscles pushes the food forwards.
- Contractions follow in waves until the food is pushed into the stomach.

Note: peristalsis occurs along the full length of the alimentary canal.

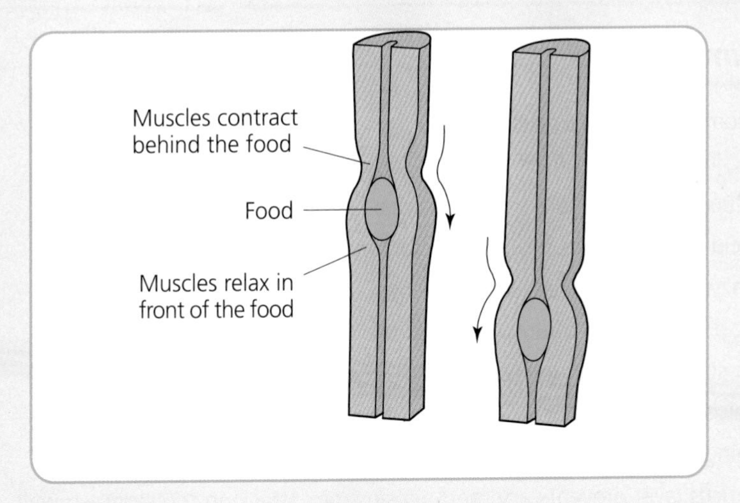

Figure 10.10

The role of the stomach in digestion

Figure 10.11 represents a section through the wall of the stomach.

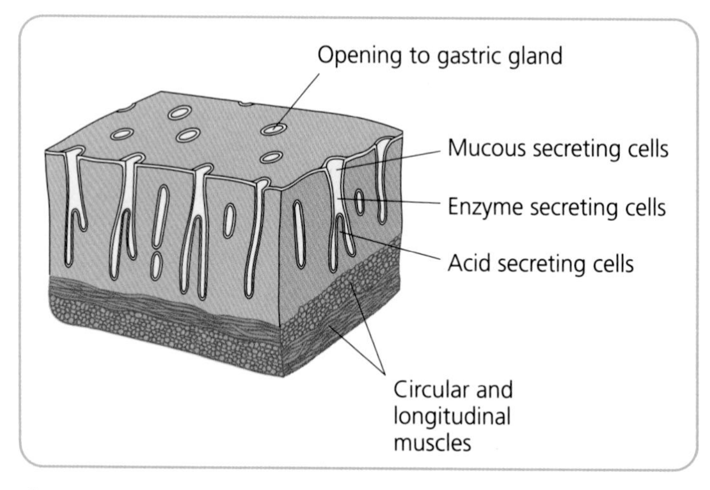

Figure 10.11

Key Points

- The stomach produces **gastric juice**.
- Gastric juice is a mixture of the secretions from **three types of cell**.
  1. Mucus-secreting cells secrete mucus.
  2. Acid-secreting cells produce hydrochloric acid.
  3. Enzyme-secreting cells secrete pepsin.
- The acid supplies the optimum pH for the action of pepsin.
- Pepsin starts the breakdown of protein to amino acids.
- Peristalsis by the **circular and longitudinal** muscles churns the food.
- Churning the food mixes it with the gastric juice.
- The mucus layer prevents acid and pepsin from attacking the stomach wall.

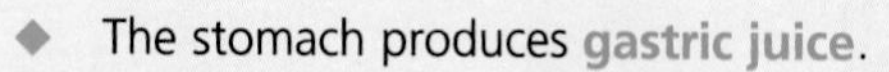
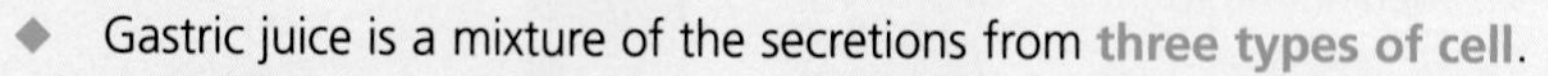

Stop Think Learn

1. Copy and complete the boxes in Figure 10.12 using information from digestion in the mouth.
2. Explain how contraction and relaxation of muscles in the oesophagus allows food to be moved from the mouth to the stomach.
3. Describe gastric juice.
4. What is the role of peristalsis in the stomach?

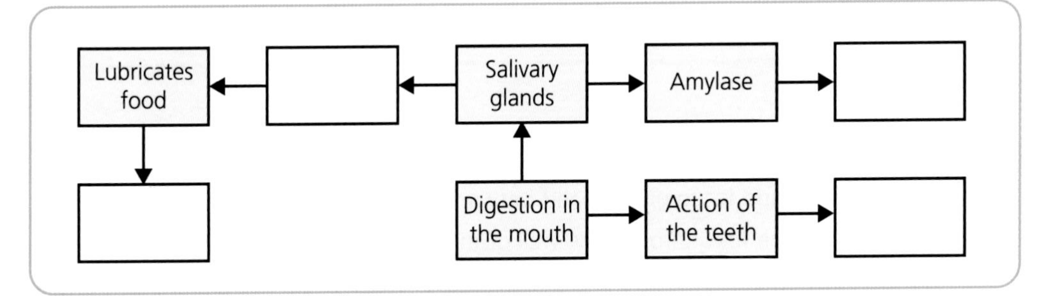

Figure 10.12

The role of the small intestine, pancreas, gall bladder and liver in digestion

Figure 10.13 shows the position of the small intestine, pancreas, gall bladder and liver.

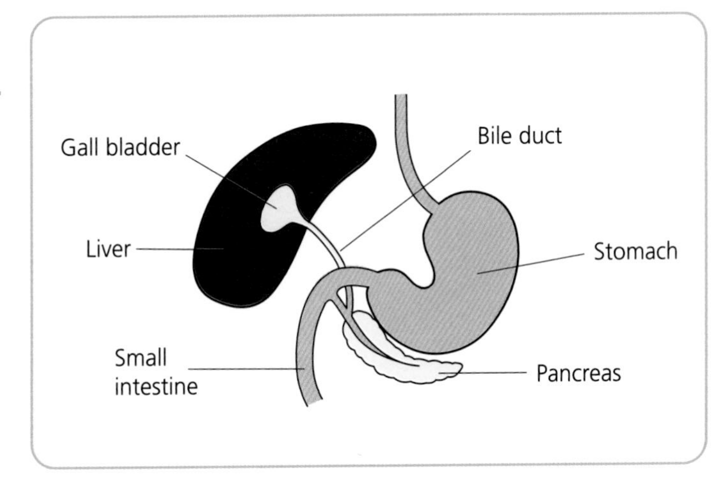

Figure 10.13

Key Points

- The **pancreas** secretes pancreatic juice into the small intestine as the food is pushed forward from the stomach.
- Pancreatic juice contains three enzymes.
  1. Amylase breaks starch down to maltose.
  2. **Trypsin** breaks protein down to amino acids.
  3. **Lipase** breaks fats down to glycerol and fatty acids.
- The liver forms **bile** from the breakdown of damaged red blood cells.
- Bile is stored in the **gall bladder**.
- Bile is secreted down the **bile duct** and mixes with the food in the small intestine as the food is pushed forward from the stomach.
- Bile breaks fats down physically into tiny droplets to form an **emulsion**.
- The droplets of fat give a large surface area for lipase enzymes to act on and this speeds up digestion.

Absorption of food through the wall of the small intestine

Key Points

- Digestion is completed in the small intestine; starch is broken down to glucose, protein to amino acids and fat to glycerol and fatty acids.
- Absorption of the food through the wall of the small intestine is by diffusion.
- The surface of the small intestine is greatly folded and protruding from the surface are microscopic finger-like projections called villi.

Figure 10.14 shows a **villus** and related structures.

- Glucose diffuses from the digested food into the blood capillaries.
- Amino acids diffuse from the digested food into the **blood capillaries of the villus**.
- Glycerol and fatty acids diffuse from the digested food into the **lacteal**.

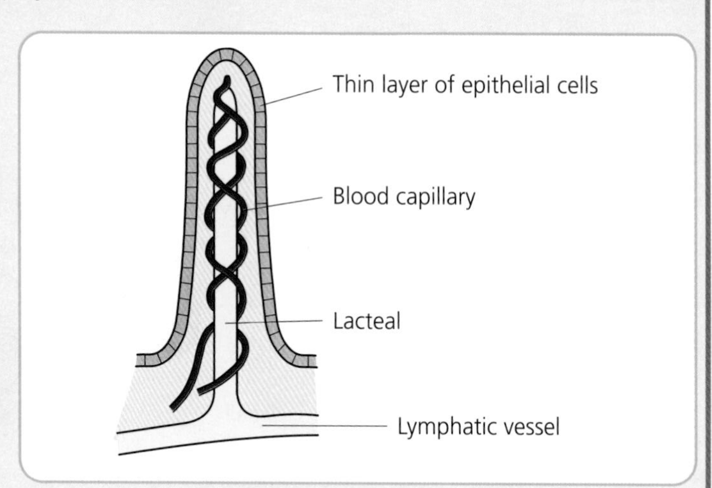

Figure 10.14

Fate of absorbed materials

Key Points

- Capillaries from the villi eventually join to form the hepatic portal vein (see diagram of the circulatory system on p. 123).
- The hepatic portal vein transports glucose and amino acids to the liver.
- Glucose passes from the liver into the bloodstream and cells of the body use glucose as a source of energy.
- In the liver, excess glucose is used to synthesise a storage carbohydrate called **glycogen**.
- Amino acids pass from the liver into the bloodstream and cells of the body use amino acids to synthesise proteins. These include enzymes and hormones.
- In the liver, excess amino acids are deaminated.
- In **deamination**, the nitrogen group of the amino acid is removed and converted to urea.
- Fats are absorbed into lacteals and pass into the lymphatic system.
- The lymphatic system links to the bloodstream.
- Fats transported into the bloodstream are used as a source of energy.

Stop Think Learn

1 Copy the diagram of the alimentary canal (see Figure 10.9). Leave out the Key.

Copy and complete the table to show the name and the function of each structure labelled from A to L. The first row has been completed for you.

Letter	Name	Function
B	Salivary gland	Produce saliva. Amylase starts the digestion of starch. Mucus lubricates the food and makes it easier to swallow.
A	Mouth	

2 Copy and complete the table to show the role of pancreatic enzymes in digestion. One of the enzymes has been named for you.

Enzyme	Substrate	Product(s)
Amylase		

3 Draw a villus and show by means of labelled arrows the diffusion pathway of glucose, amino acids, glycerol and fatty acids.

4 Produce a flowchart of the fate of absorbed materials by copying and linking the following information boxes in Figure 10.15 to the liver.

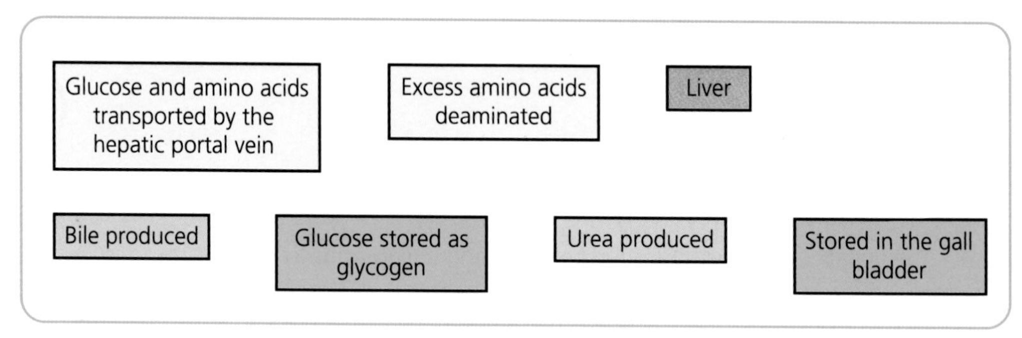

Figure 10.15

The role of the large intestine, rectum and anus in digestion

Key Points

- Peristalsis pushes undigested materials into and along the large intestine (Figure 10.16).
- Water is absorbed from the undigested material during its passage along the **large intestine**.
- The undigested material becomes more solid and forms as **faeces**.
- When the faeces is pushed into the **rectum**, muscles of the rectum contract.
- Contraction of the rectum eliminates the faeces through the **anus**.

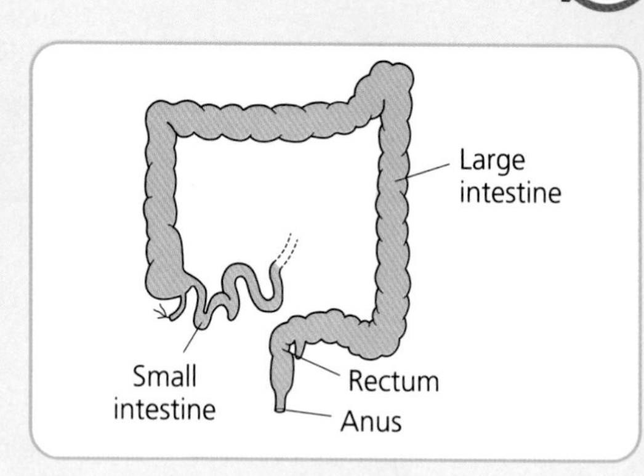

Figure 10.16

Hints and Tips

Make 'flash cards' of the words and phrases in **bold** within the text.

Remember, the pancreas secretes enzymes for the three major food classes.

The thin epithelium of the villi allows rapid diffusion (common to alveoli and capillaries). The LSA of the villi makes absorption rapid and efficient (common to alveoli and capillaries).

Questions

Examination style questions

Questions 1, 2 and 3 are based on Figure 10.17 of the human alimentary canal and related structures.

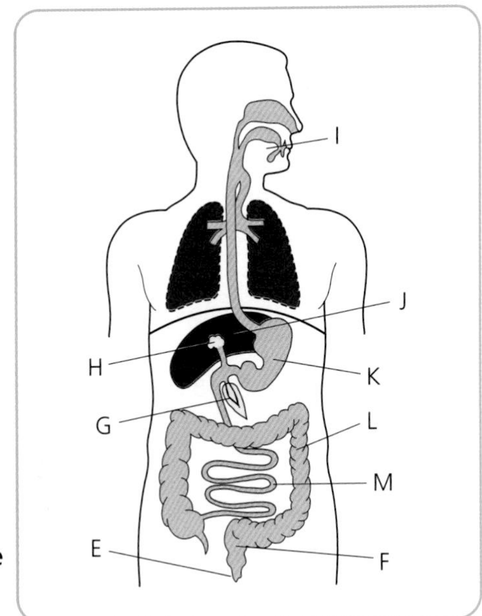

Figure 10.17

1 Two sites for the production of protein digesting enzymes are shown by:

A G and K.

B G and L.

C J and K.

D J and L.

2 Which row in the following table shows the structure that stores an emulsifying agent and the structure that absorbs most water?

	Stores emulsifying agent	Absorbs most water
A	F	J
B	H	L
C	H	F
D	J	L

3 Which row in the following table shows a gland that produces a chemical to aid swallowing and a gland that produces both amylase and lipase?

	Aids swallowing	Produces amylase and lipase
A	G	M
B	K	M
C	I	K
D	I	G

4 Decide whether each of the following statements is **TRUE** or **FALSE** and **tick the appropriate box** in the following table.

If you decide the statement is **FALSE**, you should then write the **correct**

Questions continued

word(s) in the right-hand box to replace the word(s) underlined in the statement. *(3)*

Statements	True	False	Correct word
Vitamins in the human body are essential for good health.			
An element present in amino acids that is absent from carbohydrate and fats is oxygen.			
Biuret's test is used to show that fat is present in a food.			

5 Figure 10.18 represents a section through a villus from the small intestine.

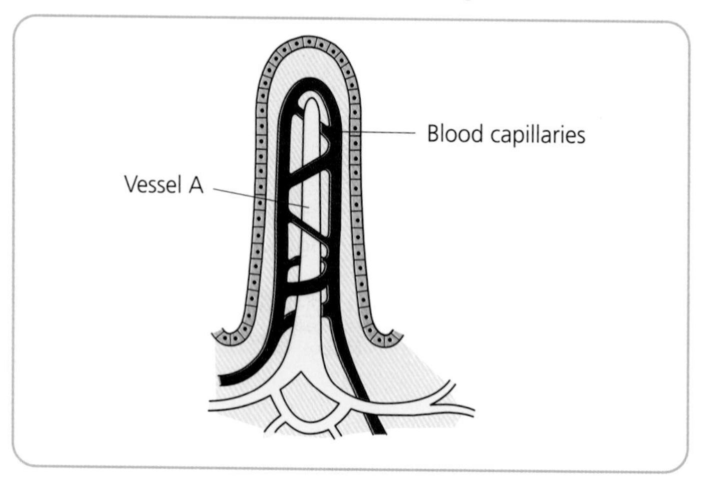

Figure 10.18

a) Vessel A is involved in the absorption of the products of the digestion of one of the food groups.

i) Name vessel A. *(1)*

ii) Name the products that are absorbed. *(1)*

b) i) Name the organ that stores excess glucose and state the form in which it is stored. *(1 + 1)*

ii) What use does the body make of the amino acids absorbed? *(1)*

iii) Name the waste product formed from deamination of amino acids in the liver. *(1)*

6 Describe the chemical composition and structure of carbohydrates and proteins. *(5)*

Answers

Answers to examination type questions with commentary

1 You should know that the stomach and pancreas produce protein-digesting enzymes pepsin and trypsin. These are shown by the letters K and G. **Answer = A**

2 Emulsifying agent is bile and it is stored in the gall bladder. Most water is absorbed in the large intestine. These are shown by the letters H and L. **Answer = B**

3 Swallowing is aided by mucus produced in saliva. Lipase and amylase are produced by the pancreas. These are shown by the letters I and G. **Answer = D**

4 You should know that vitamins are related to good health. **Answer = TRUE**

Carbon, hydrogen and oxygen are common to all food classes. Protein also contains nitrogen. **Answer = FALSE** Nitrogen

Biuret's test is for protein. A translucent spot shows that fats are present.

Answer = False Translucent spot. 1 mark for each correct answer.

5 a) i) You should know that this is a lacteal.

ii) Absorbs glycerol and fatty acids. Note: question uses pleural so more than one product must be given.

b) i) Liver stores excess glucose as glycogen.

ii) You should know that amino acids are the sub-units for the synthesis of proteins such as enzymes and hormones.

iii) Deamination removes the nitrogen part from amino acids and urea is formed.

6 Describe the chemical composition and structure of carbohydrates and proteins.

In extended writing questions, you **must** identify all the parts that make up the question. These are:

- Describe the chemical composition of carbohydrates.
- Describe the chemical composition of proteins.
- Describe the chemical structure of carbohydrates.
- Describe the chemical structure of proteins.

Marking Instructions

A1 Carbohydrates and proteins contain carbon, hydrogen and oxygen.

A2 Proteins also contain nitrogen.

Maximum 2 marks.

B1 Basic units of carbohydrates are sugars/glucose.

B2 Glucose molecules bond to form chains of starch/cellulose/glycogen.

B3 Twenty different types of amino acids used.

B4 Bond to form protein.

B5 Structure depends on a sequence of amino acids.

Maximum 4 marks. Overall maximum 5 marks.

Chapter 11

CONTROL OF THE INTERNAL ENVIRONMENT

Structure of the Human Urinary System

Figure 11.1 shows the kidneys and related structures.

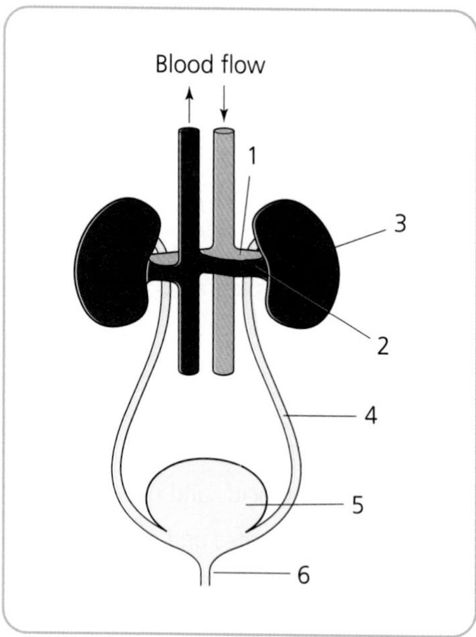

Figure 11.1

Key

1. **Renal arteries** – transport blood to the kidneys.
2. **Renal veins** – transport blood away from the kidneys.
3. **Kidneys** – filter blood and produce urine.
4. **Ureters** – transport urine to the bladder.
5. **Bladder** – stores urine.
6. **Urethra** – urine is passed out of the body from the bladder.

Use the key to identify:
1. structures associated with the kidney;
and 2. the function of each structure.

Role of the Kidney

The kidney has two main functions: **osmoregulation** and **excretion of nitrogenous waste in the urine**.

Osmoregulation

Keeping the water concentration of the blood and tissue fluids of an organism constant is called osmoregulation and is carried out by the kidneys.

Without osmoregulation cells of the body could take in water and swell and burst or lose water and shrink.

In osmoregulation water loss must equal water gain.

The table shows the water balance over a 24-hour period for a healthy human.

Water gain	cm^3	Water loss	cm^3
In drink	1200	In sweat	400
In food	950	In breath	400
In metabolic water	350	In faeces	100
		In urine	1600
Total	2500	Total	2500

Production and removal of urea

Excess amino acids are deaminated in the liver. **Deamination** is the removal of the nitrogen containing part of the amino acid.

Urea is a nitrogenous waste formed from the nitrogen part of the amino acid.

Urea is transported from the liver in the blood and enters the kidneys through the renal arteries.

Urea is removed from the blood by the kidneys and is passed out of the body in the urine.

Structure and function of the kidney

The working unit of the kidney is the nephron. Each kidney contains around one million nephrons.

Figure 11.2 shows the structure of a nephron.

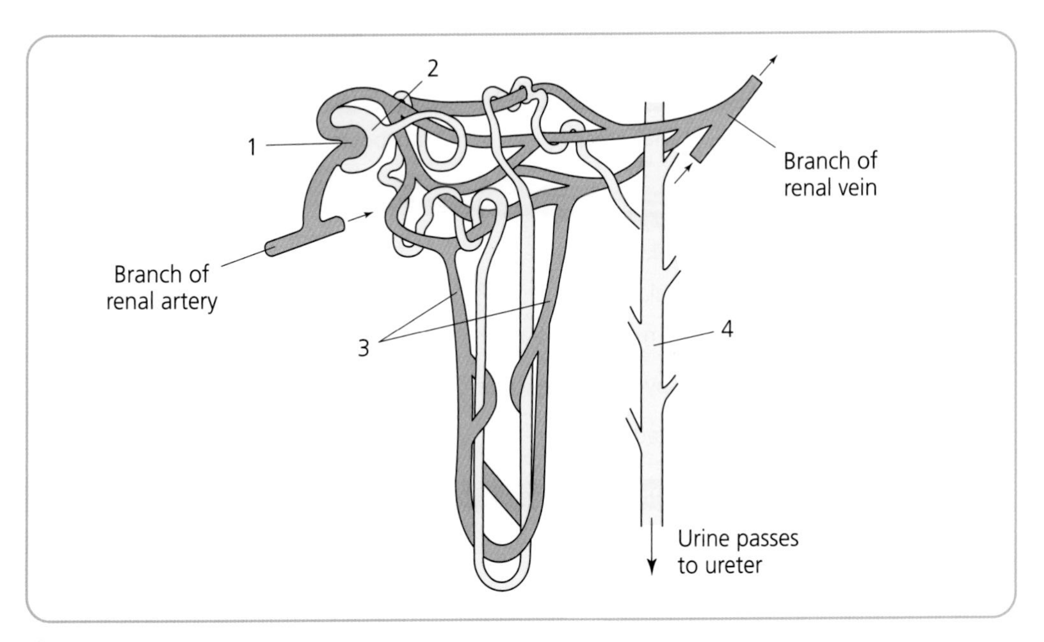

Figure 11.2

Use the key on the following page to identify:
1. structures associated with the nephron; and 2. the function of each structure.

Key

1. **Glomerulus** – a knot of capillaries formed from a branch of the renal artery. Some of the liquid part of the blood filters through the walls of the glomerulus.
2. **Bowman's capsule** – the filtrate passes into Bowman's capsule. The filtrate is made up mainly of water, glucose, salts and urea.
3. **Capillaries** – reabsorb all of the glucose, most of the salts and much of the water as the fluid passes down the tubule.
4. **Collecting ducts** – the final reabsorption of water takes place here.

The volume of water reabsorbed depends on the water concentration of the blood and is controlled through the hormone ADH (antidiuretic hormone).

The fluid that leaves the collecting ducts and passes to the ureter is urine.

Urine contains water, excess salts and urea.

Stop Think Learn

1 Copy the diagram of the kidney and associated vessels (Figure 11.1). Leave out the Key.

Copy and complete this table by naming each structure and giving its function.

The renal artery has been completed for you.

Number	Structure	Function
1	Renal artery	Transports blood to the kidneys
2		

2 State two functions of the kidneys.

3 What is meant by water balance?

4 Copy the diagram of the structure of a nephron (Figure 11.2). Leave out the Key.

Copy and complete this table by naming each structure and giving its function.

The first row has been completed for you.

Number	Name	Function
1	Glomerulus	Filtration of the fluid part of the blood
2		

Negative Feedback Control by Antidiuretic Hormone (ADH)

Figure 11.3 outlines part of the regulation of water concentration of the blood.

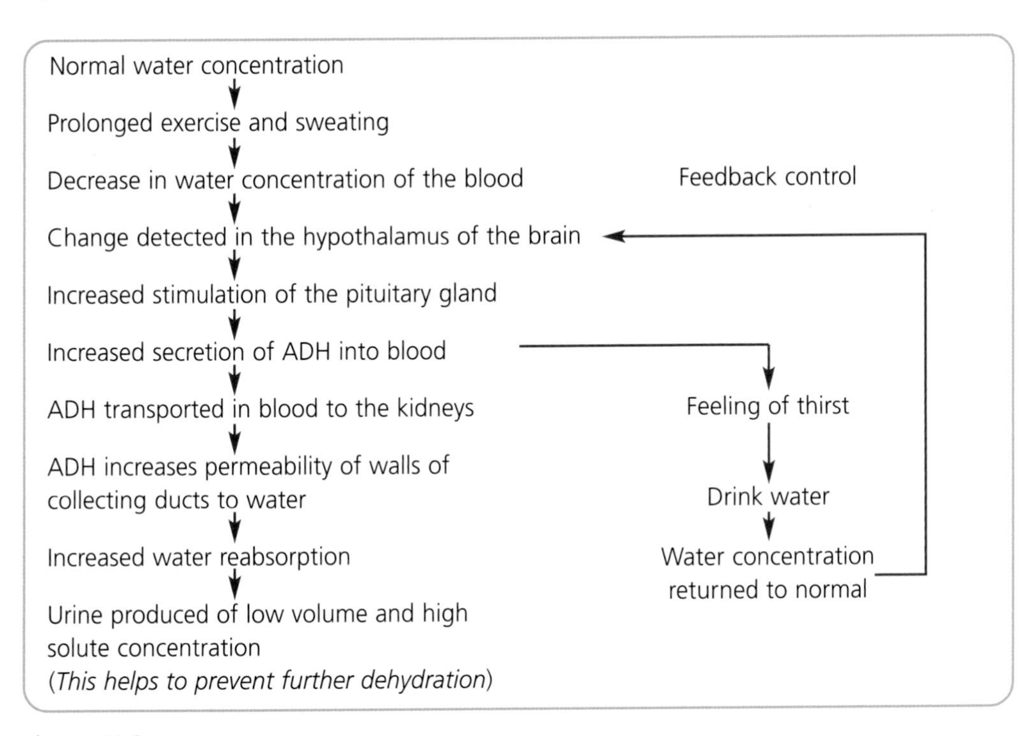

Figure 11.3

The feedback control outlined is described as **negative feedback** control. This is when the return to normal is detected in the correction centre (*hypothalamus*) and the corrective mechanism (increased ADH) is switched off. Negative in this context means to negate or stop. The corrective mechanism that is working is stopped.

Stop Think Learn

Draw out a similar flowchart to show what happens when the water concentration of the blood increases as a result of drinking too much water when not thirsty.

Osmoregulation in Marine and Freshwater Bony Fish

One of the **problems of marine bony fish** is that their tissues are hypotonic to sea water.

Hypotonic means a lower solute concentration, which means a higher water concentration. (Draw out the water concentration triangle.)

Water is lost continuously by osmosis into the surrounding sea water. Figure 11.4 shows how the problem is overcome.

1 Fish drink sea water.
2 Additional salts taken in with the sea water are removed by the gills.
3 To retain water, the kidneys produce a small volume of concentrated urine.

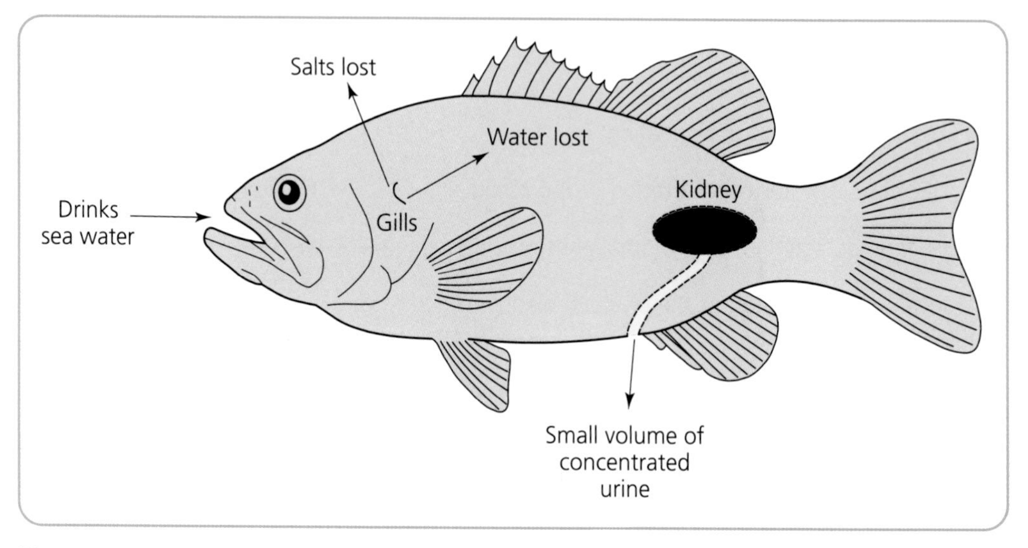

Figure 11.4

One of the **problems of freshwater bony fish** is that their tissues are hypertonic to sea water and water enters continuously by osmosis from the surrounding freshwater. (Hypertonic means a higher solute concentration, which means a lower water concentration.)

Tissue cells are in danger of swelling. Figure 11.5 shows how the problem is overcome.

1 To remove water, kidneys produce a large volume of dilute urine.
2 To overcome salt loses in the urine, salts are taken in at the gills.

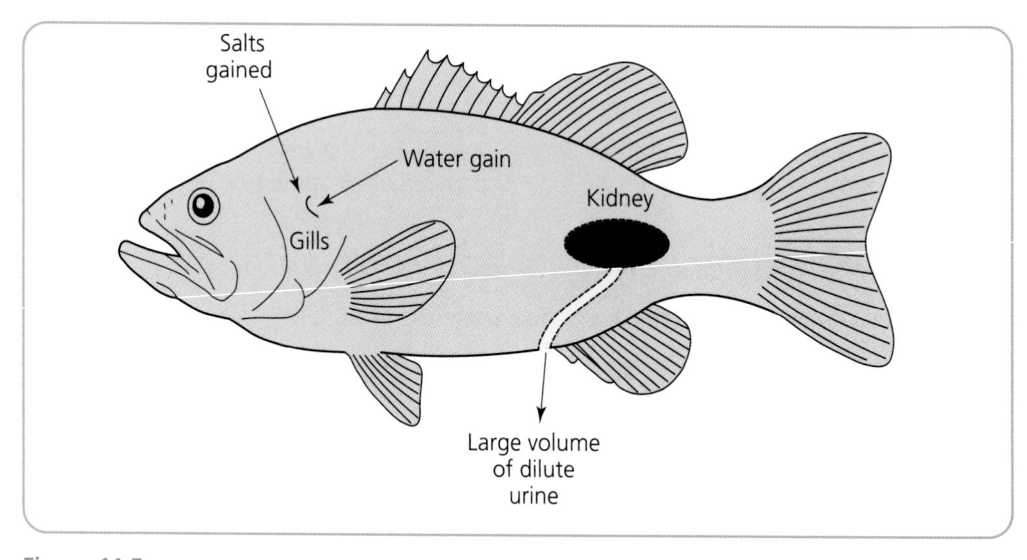

Figure 11.5

Hints and Tips

Make 'flash cards' for the words and phrases in **bold**.

For every diagram, be able to name the structures and give their functions.

By understanding the problems in osmoregulation for one fish then you should understand them for both as they are opposites. To make something 'Fresh' you put it into water so it takes up water. In freshwater fish water comes in to keep the fish 'fresh'.

In descriptions of urine formation in the nephron, use 'reabsorption' not 'absorption'. This is to emphasise that it has been filtered out and it is now being returned.

Questions

Examination style questions

1 Figure 11.6 shows the structure of the human urinary system.

Which of the following is correct?

A X is the renal artery, Y is the renal vein and Z is the urethra.

B X is the renal vein, Y is the renal artery and Z is the urethra.

C X is the renal artery, Y is the renal vein and Z is the ureter.

D X is the renal vein, Y is the renal artery and Z is the ureter.

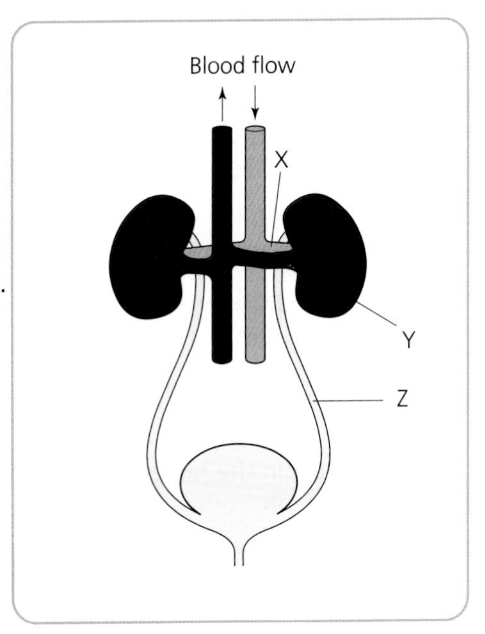

Figure 11.6

2 In the human brain the centre responsible for osmoregulation is found in the:

A pituitary.

B cerebellum.

C cerebrum.

D hypothalamus.

Questions continued

3 Drinking a large volume of water when not thirsty would result in:

A a decreased secretion of ADH and an increased volume of urine.

B an increased secretion of ADH and a decreased volume of urine.

C an increased secretion of ADH and an increased volume of urine.

D an decreased secretion of ADH and a decreased volume of urine.

4 The table below shows the concentration of chemical substances in the blood plasma, the glomerular filtrate and the urine of a healthy human.

Substance	Plasma (%)	Glomerular filtrate (%)	Urine (%)
Water	90.0	90.0	95.0
Protein	8.0	0.0	0.0
Glucose	0.10	0.10	0.0
Urea	0.030	0.030	2.01

a) Explain the difference in the composition of the blood plasma and the glomerular filtrate. *(1)*

b) Explain the difference in the glucose concentrations of the glomerular filtrate and the urine. *(1)*

c) By how many times is the urea concentration of the urine greater than that of the plasma? *(1)*

5 a) Describe the problem that marine bony fish face in their environment that makes them have to osmoregulate. *(2)*

b) In the table, tick (✓) three lines which show features of osmoregulation in freshwater bony fish. *(2)*

Features	Tick (✓)
Excrete excess salts	
Take up salts	
Produce a small volume of urine	
Produce a large volume of urine	
Produce a concentrated urine	
Produce a dilute urine	

6 Describe the role of the nephron and collecting ducts in the production of urine. *(5)*

Answers

Answers to examination type questions with commentary

1 The arrow in the blood vessel connected to the vessel labelled Y is taking blood away and for X it is carrying blood to the kidney. X is the artery and Y the vein.

You should know that Z is the ureter. **Answer = C**

2 You should know that osmoreceptors detect changes in water concentration of the blood and that these are in the hypothalamus. **Answer = D**

3 This would increase the water concentration of the blood. ADH secretion would decrease. Less water would be reabsorbed from the collecting ducts and as a result urine volume would increase. **Answer = A**

4 a) You have to notice that the only difference in the compositions of the plasma and filtrate are the proteins. The only reason for something not filtering is size.

Answer = The protein molecules are too large to be filtered.

b) Between the glomerulus and urine formation the fluid has passed through the tubules. You should know that reabsorption takes place in the tubules.

Answer = Glucose is reabsorbed as the fluid passes along the tubule.

c) By how many times: divide one number by the other. In urine = 2.01 and in plasma = 0.030. Answer 2.01 divided by 0.030 = 67 times.

These calculations can be made easier by removing the fractions. Start with the smallest number. If the point is moved twice 0.03 becomes 3. If you move the point in one number you must do the same in the other = 2.01 becomes 201.

This is an easier calculation: 201 divided by 3.

5 a) Their tissues have a higher water concentration than sea water *1 mark*

and as a result water is constantly lost by osmosis. *1 mark*

b) Tissues of freshwater fish have a lower water concentration than surroundings.

Water continuously enters by osmosis. Tick Large volume, dilute urine and take up salts

3 correct = 2 marks, 2 correct = 1 mark, 1 correct = 0

6 Describe the role of the nephron and collecting ducts in the production of urine.

In extended writing questions you **must** identify all the parts that make up the question. These are:

- Describe the role of the nephron in the production of urine.
- Describe the role of the collecting ducts in the production of urine.

Marking instructions

A1 Filtration of blood at the glomerulus.

A2 Filtrate passes into Bowman's capsule.

Answers continued

A3 Reabsorption occurs in the tubule.

A4 Reabsorption into blood capillaries.

A5 All the glucose, some salts and some water are reabsorbed.

A6 Urea not reabsorbed.

Maximum of 3 marks.

B1 Collecting ducts reabsorb water.

B2 Volume of water reabsorbed depends on water concentration of the blood.

B3 Fluid that passes from the collecting ducts is urine.

Maximum 2 marks.

Overall maximum = 5 marks.

Chapter 12

CIRCULATION AND GAS EXCHANGE

Structure and Function of the Heart

Structure of the heart related to its function as a muscular pump

Figure 12.1 shows the structure of the heart and associated vessels.

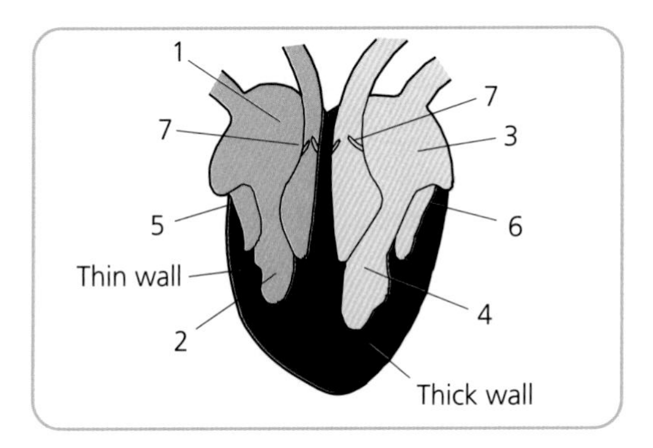

Figure 12.1

Use the key to identify:
1. the structures of the heart,
2. the function of each structure.

Key

1 **Right atrium** – collects blood from all parts of the body other than the lungs and pumps blood into the right ventricle.
2 **Right ventricle** – contracts and generates pressure that pumps blood to and through the lungs to the left atrium.
3 **Left atrium** – collects blood from the lungs and pumps blood into the left ventricle.
4 **Left ventricle** – contracts and generates pressure that pumps blood to and through all parts of the body, other than the lungs, to the right atrium.
The left ventricle has to pump blood further than the right ventricle. To create greater pressure more muscle is required. The wall is thicker than that of the right ventricle.
5 **Tricuspid valve** – prevents blood flowing back into the right atrium when the right ventricle contracts.
6 **Bicuspid valve** – prevents blood flowing back into the left atrium when the left ventricle contracts.
7 **Semi-lunar valves** – prevent blood flowing back from the arteries into the ventricles when the ventricles relax.

Blood vessels

Blood leaves the heart in arteries, flows through capillaries and is returned to the heart in veins.

Figure 12.2 show the structure of an artery, vein and capillaries.

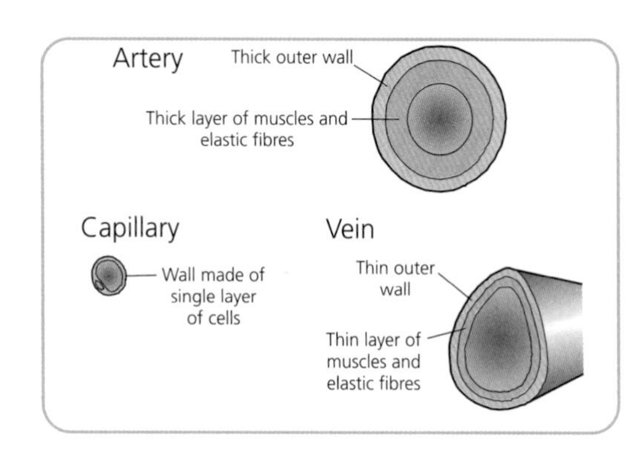

Figure 12.2

Artery

Arteries have to absorb the pressure generated by contraction of the ventricles.

Elastic tissue in the artery wall stretches and recoils as the blood passes through. This is felt as a **pulse**. Contraction and relaxation of the **muscle tissue** is used to alter the diameter of an artery and this allows blood flow through an artery to be controlled.

Capillaries (see Figure 12.5)

Capillaries **form as a network** within all the tissues thus every cell of the body is close to a capillary.

Exchange of food, gases and waste products takes place by diffusion through the walls of the capillaries. The exchange is between the fluid that bathes the cells (tissue fluid) and the blood in the capillary.

Capillaries are **thin walled**, as they are only one cell thick.

The thin wall allows more rapid exchange between blood and tissue fluid.

Capillaries have a **large surface area** and this ensures that the rate of exchange between blood and tissue fluid is more rapid and efficient.

Veins

The walls of veins have **less elastic and muscle tissues** than arteries.

Veins do not have to absorb the high pressure generated by contraction of the ventricles. Veins have **valves**. The pressure generated by contraction of muscles forces blood along the veins and through the valves.

If pressure forces blood to flow away from the heart the valves close over.

Closing of the valves prevents back flow of blood.

Figure 12.3 shows the position of the coronary arteries and other blood vessels associated with the heart.

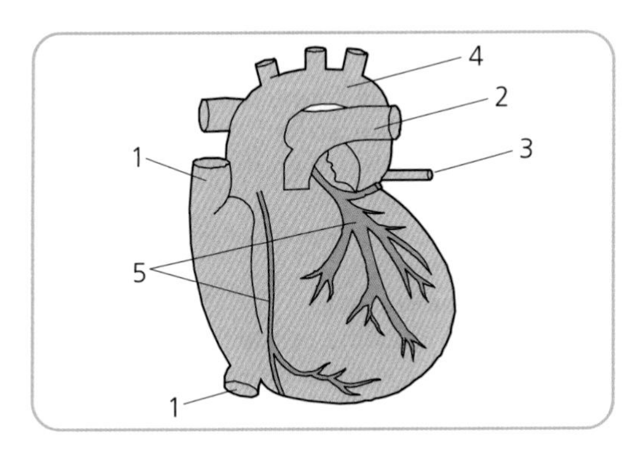

Use the key to identify:

1. the blood vessels,
2. the function of each blood vessel.

Figure 12.3

Key

1 Vena cava – transports blood from all parts of the body, other than the lungs, to the right atrium.
2 Pulmonary artery – transports blood to the lungs from the right ventricle.
3 Pulmonary vein – transports blood from the lungs to the left atrium.
4 Aorta – transports blood from the left ventricle to all parts of the body, other than the lungs.
5 Coronary arteries – transport blood to the heart muscles.

If a coronary artery is blocked, blood no longer reaches the muscles in that area of the heart. Starved of food and oxygen, the heart muscles no longer respire. The heart muscles in the area affected stop contracting. This is a coronary heart attack.

Circulation of Blood

Figure 12.4 shows major blood vessels of the circulatory system.

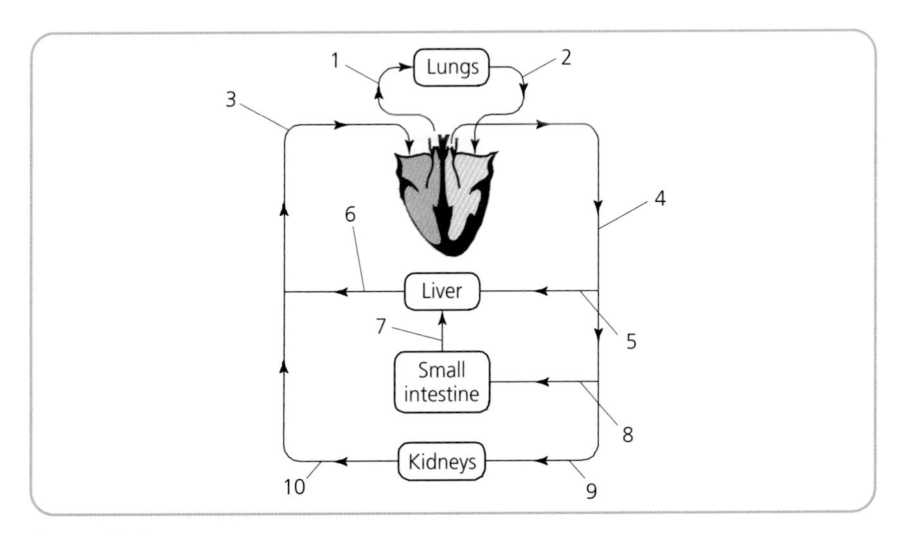

Figure 12.4

Use the key on the following page to identify:
1. the position of each blood vessel,
2. the function of each blood vessel.

Key

1 Pulmonary artery – transports blood from the right ventricle to the lungs.
2 Pulmonary vein – transports blood from the lungs to the left atrium.
3 Vena cava – transports blood from all parts of the body other than the lungs to the right atrium.
4 Aorta – transports blood from the left ventricle to all parts of the body other than the lungs.
5 Hepatic artery – transports blood from the heart to the liver.
6 Hepatic vein – transports blood from the liver to the vena cava.
7 Hepatic portal vein – transports blood from the small intestine to the liver.
8 Mesenteric artery – transports blood from the heart to the small intestine.
9 Renal artery – transports blood from the heart to the kidneys.
10 Renal vein – transports blood from the kidneys to the vena cava.

Stop Think Learn

1 Copy the diagram of the heart and associated vessels, leaving out the key.

2 Copy and complete this table by naming each numbered structure in Figure 12.1 and giving its function. The right atrium has been completed for you.

Number	Structure	Function
1	Right atrium	Collects blood from all parts of the body other than the lungs and pumps blood into the right ventricle
2		

Check your answers from the information in the text.

3 Copy and complete the flowchart to show the path of blood flow through the heart.

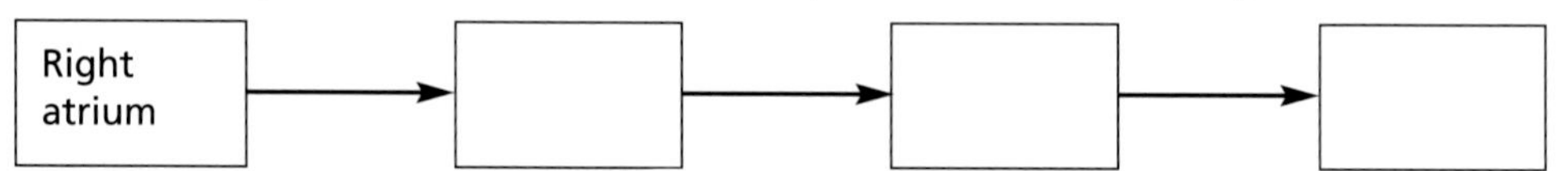

4 Copy the diagram of the major blood vessels of the circulatory system, leaving out the key. Copy and complete the table below by naming each numbered blood vessel and giving its function. The pulmonary artery has been completed for you.

Number	Blood vessel	Function
1	Pulmonary artery	Transports blood from the right ventricle to the lungs
2		

Check your answers from information in the text.

5 The sentences refer to the structure of blood vessels. Copy and complete the blanks.

Pressure from contraction of the ventricles causes ____________ tissue in an ________ to stretch and recoil. As the blood flows this is felt as a _____________.

Muscles in the _____ of the arteries can contract to _______ blood flow.

Veins have _____________ to prevent ___________ of blood.

Capillaries are ______ walled and present a large_______ ________ for exchange.

Every cell in the body is ________ to a capillary due to the fact that capillaries form a _____________.

Composition and Functions of Blood

Function of red blood cells and plasma in the transport of respiratory gases and food

Plasma is the fluid part of the blood and blood cells are suspended in the plasma. Oxygen is carried by red blood cells (RBC) (Figure 12.5).

Carbon dioxide is carried dissolved in the plasma and some is carried in RBC.There is a limit to the concentration of carbon dioxide that can dissolve in the plasma. As carbon dioxide dissolves, the acidity of the blood increases and as the acidity increases less carbon dioxide can dissolve.

Soluble food (glucose and amino acids) is carried in the plasma.

Food is carried to the cells and is used in respiration and for the functions of the cell.

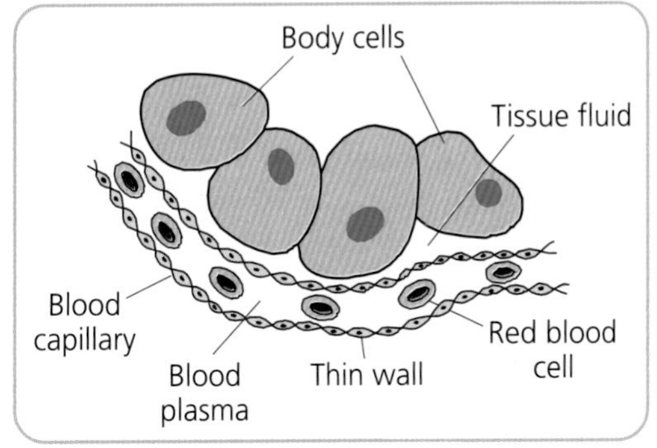

Figure 12.5 A capillary and surrounding tissue

Function of haemoglobin in the transport of oxygen

Haemoglobin combines with oxygen when the oxygen concentration is high.

Haemoglobin combines with oxygen to form oxyhaemoglobin.

Oxyhaemoglobin releases oxygen when the oxygen concentration is low.

Oxygen concentration is high in the lungs and low in the tissues. The low concentration is due to aerobic respiration in the tissues.

The equation outlines the reversible reaction of haemoglobin with oxygen.

$$\text{Haemoglobin} + \text{oxygen} \underset{\text{Low oxygen concentration in the tissues}}{\overset{\text{High oxygen concentration in lungs}}{\rightleftharpoons}} \text{oxyhaemoglobin}$$

Functions of macrophages and lymphocytes in defence

Figure 12.6 shows stages in **phagocytosis** by a **macrophage**.

Stage 1. The cell membrane of the macrophage surrounds the bacterium.

Stage 2. The bacterium is engulfed by the macrophage.

Stage 3. Digestive enzymes are added.

Stage 4. The enzymes digest the bacterium.

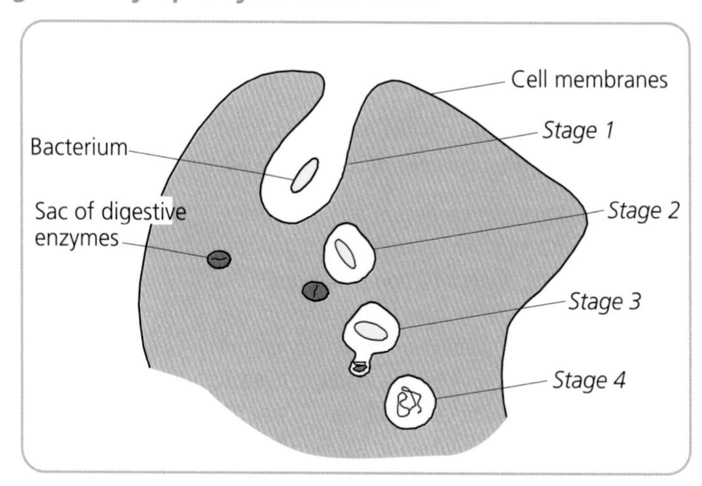

Figure 12.6

Figure 12.7 shows stages in **antibody production** by a **lymphocyte**.

Stage 1. The presence of bacteria is detected by a lymphocyte.

Stage 2. The lymphocyte divides rapidly to produce many cells.

Stage 3. The lymphocytes produce antibodies.

Stage 4. The antibodies cause the bacteria to be destroyed.

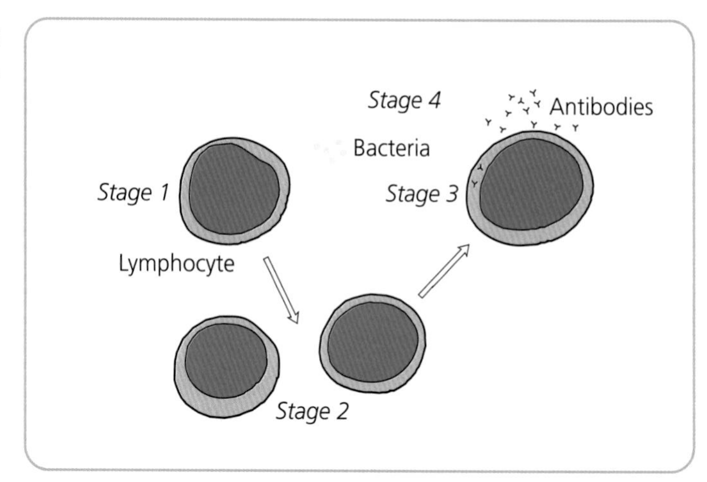

Figure 12.7

Each type of virus and bacterium is recognised by a particular lymphocyte.

Lymphocytes produce antibodies that destroy only one particular bacterium or virus.

The antibodies are said to be specific.

Stop Think Learn

1 Copy and complete this table to show transport in the blood. Oxygen has been completed for you.

Substance	How transported
Oxygen	Transported in red blood cells as oxyhaemoglobin

2 Describe the reversible reaction of haemoglobin with oxygen.

3 Copy and complete the flowchart to show stages in phagocytosis.

The first stage has been completed for you.

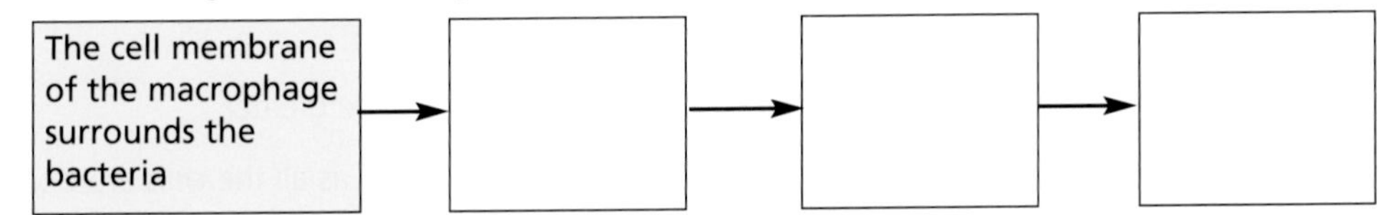

4 Complete the blanks in the sentences that refer to antibody production.

The cells that produce antibodies are called ________________.

Antibodies are described as being ____________ because each type of antibody destroys only the one type of _____________ or virus.

Structure and Function of the Lungs in Gas Exchange

Internal structure of lungs

Figure 12.8 shows the structure of the lungs.

Use the key to identify the structures.

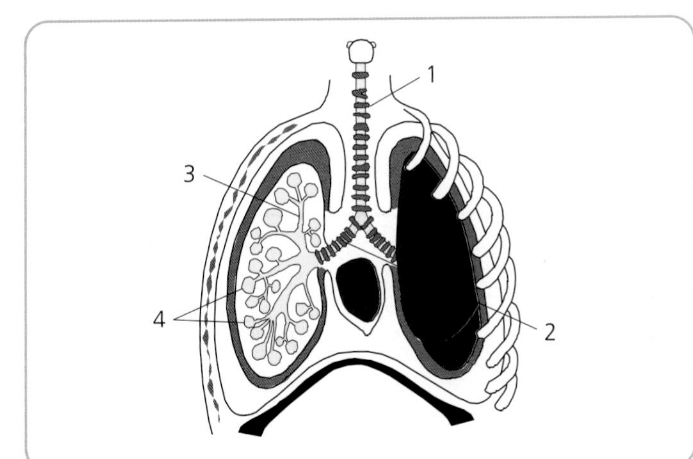

Figure 12.8

Key

1 **Trachea**
2 **Bronchus** (Pleural bronchi)
3 **Bronchioles**
4 **Alveoli** (air sacs)

Figure 12.9 shows an alveolus together with a blood capillary.

Features of alveoli allow efficient gas exchange.

The **moist surface** of the alveolus allows oxygen from the air to dissolve.

Oxygen concentration in the moisture is higher than in the blood in the capillary.

Carbon dioxide concentration is higher in the blood than in the air space of the alveolus.

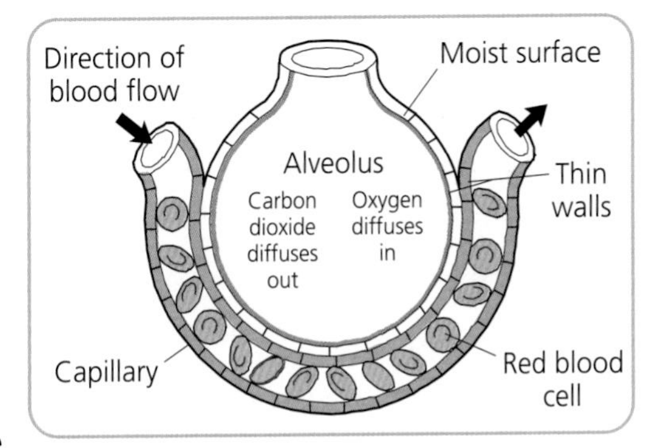

Figure 12.9

Oxygen diffuses from the moisture layer through the walls of the alveolus and capillary into the blood.

Carbon dioxide diffuses from the blood through the walls of the capillary and the alveolus and into the air space of the alveolus.

Thin walls allow gas exchange by diffusion to be faster.

The **large surface area** of the alveolus allows gas exchange to be greater.

A **rich capillary network** allows gas exchange to be more rapid as all the cells are close to a capillary.

Stop Think learn

1. Copy the diagram of the lungs, leaving out the key.
2. Copy and complete this table by naming each numbered structure.

Number	Structure
1	

3. Copy and complete this table to show how structural features of alveoli make them efficient at gas exchange.

   Moist surface has been completed for you.

Structural feature	Role in efficiency of gas exchange
Moist surface	Allows oxygen from the air in the alveolus to dissolve

Check your answers from the information in the text.

Hints and Tips

Make 'flash cards' of the words and phrases in **bold** in the text.

Starting from the right atrium you can use RAV – LAV to remember the path of blood through the heart. Right atrium – right ventricle – left atrium – left ventricle.

Tricuspid has an R in it. R = right: the valve lies between the right atrium and the right ventricle.

The pulmonary artery is the only artery that carries deoxygenated blood. The pulmonary vein is the only vein that carries oxygenated blood.

The liver has three blood vessels. As well as the hepatic artery and vein, the hepatic portal vein transports the products of digestion from the small intestine.

In all areas of exchange in the body three key points are:

THIN – LARGE SURFACE AREA – RICH CAPILLARY SUPPLY

In questions that relate to explanations of thin walls, etc. you must use comparative terms such as greater, faster, more efficient, etc.

The heart cycle includes contraction of the atria, then the ventricles and then relaxation between contractions.

Questions

Examination style questions

1 In Figure 12.10 which part (A, B, C or D) shows the structure of an artery?

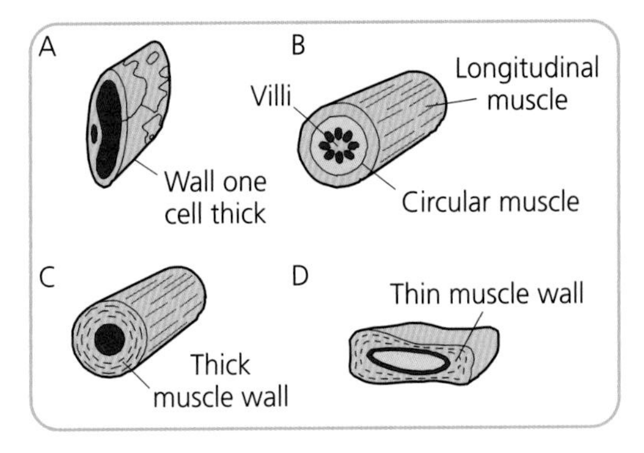

Figure 12.10

2 Figure 12.11 shows the time taken for different stages within three heart cycles when a human subject was at rest.

The total time taken for each heart cycle is

A 2.4 seconds.

B 0.8 seconds.

C 0.5 seconds.

D 0.08 seconds.

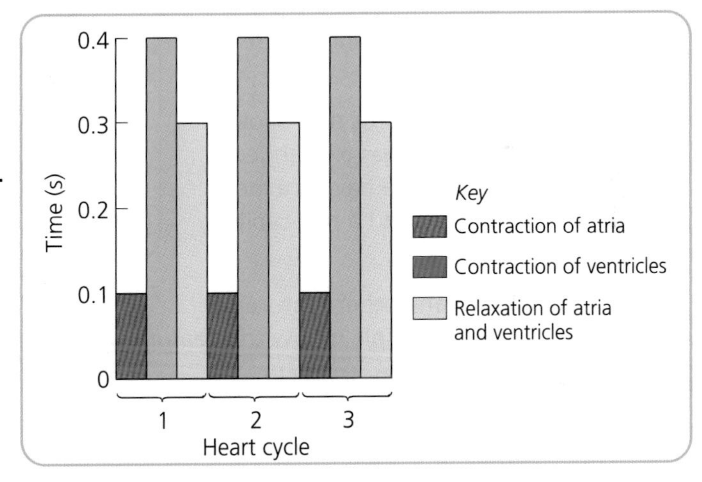

Figure 12.11

3 Figure 12.12 represents the human heart and part of the circulatory system.

a) Name the chamber of the heart that would first receive nicotine from inhaled cigarette smoke. *(1)*

b) Name blood vessel A and organ Y. *(2)*

c) State the position of the tricuspid valve. *(1)*

Questions continued

d) What evidence from the diagram supports the statement that the left ventricle creates a greater pressure than the right ventricle? *(1)*

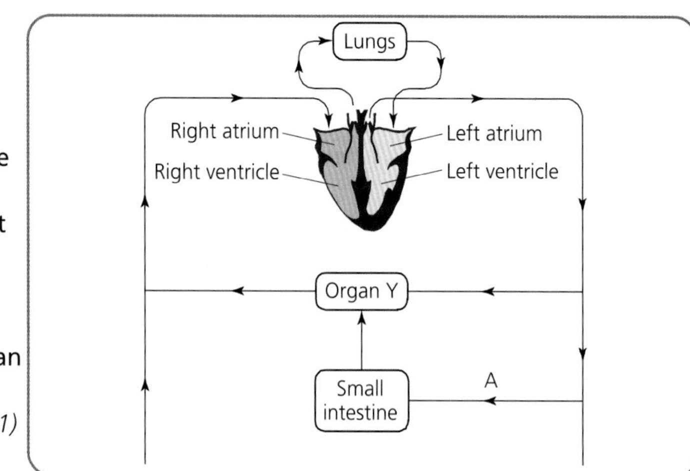

Figure 12.12

4 Figure 12.13 represents an alveolus together with a blood capillary.

a) Explain why oxygen concentration in the capillary is higher at B than at A. *(2)*

b) Two features that allow efficient gas exchange at alveoli are having a moist surface and a rich capillary supply.

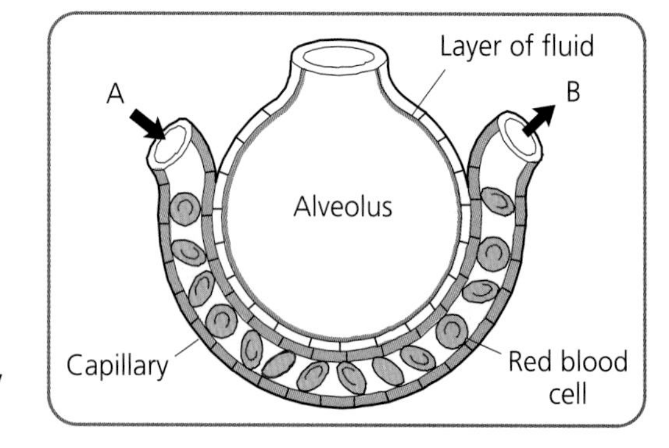

Figure 12.13

Identify one other feature of alveoli and explain how this feature is related to efficient gas exchange. *(1 + 1)*

5 Describe the function of blood in the transport of respiratory gases and food. *(5)*

Answers

Answers to examination type questions with commentary

1 You should know that arteries have thicker walls than veins and that capillaries are only one-cell thick. Villi are found in the small intestine, not in arteries. **Answer = C**

2 You have to add up the value of the bars for contraction of the atria and ventricles plus the relaxation = 0.1 + 0.4 + 0.3 Total = 0.8. **Answer = B**

Answers continued

3 a) Nicotine enters at the lungs. Blood returns to the heart from the lungs by the pulmonary vein. The pulmonary vein enters at the **left atrium** of the heart.

b) Vessel A is taking blood from the heart – it is an artery – to the small intestine.

It is the **mesenteric artery**. Organ Y is receiving blood from the small intestine. You should know that this is the **liver**.

c) It has an R in it. R for right. It lies between the right atrium and the right ventricle.

d) You should know that the pressure generated is related to the thickness of the muscular walls of the ventricles. Evidence is that muscle wall is thicker.

4 a) Question is for 2 Marks so two good points are required.

It is because the blood flows from A to B, *(1st mark)*

and as it flows oxygen diffuses in along the length of the capillary. *(2nd mark)*

b) You should know that other features are large surface area or thin wall. *(1st mark)*

If you answered **LSA** then the explanation is – greater gas exchange.

If you answered **thin wall** then the explanation is – rate of diffusion is faster. *(2nd mark)*

5 Describe the function of blood in the transport of respiratory gases and food.

In extended writing questions you must identify all the parts that make up the question. These are:

- Describe the function of blood in the transport of respiratory gases.
- Describe the function of blood in the transport of food.

A1 Oxygen carried in red blood cells.

A2 Carbon dioxide carried in red blood cells.

A3 Carbon dioxide carried dissolved in the plasma.

A4 Concentration of carbon dioxide carried in the plasma is limited by the increase in acidity as a result of carbon dioxide dissolving.

A5 Haemoglobin combines with oxygen in lungs at a high oxygen level.

A6 Oxygen released by oxyhaemoglobin in the tissues at a low oxygen level.

Maximum of 4 marks.

B1 Food carried dissolved in the plasma.

B2 Food carried to the cells of the body for respiration and growth.

Maximum of 1 mark. Maximum total = 5 marks.

Chapter 13

SENSORY MECHANISMS AND PROCESSING OF INFORMATION

Structure and Function of the Brain

Figure 13.1 shows a section of the brain and 13.2 shows a surface view of the brain.

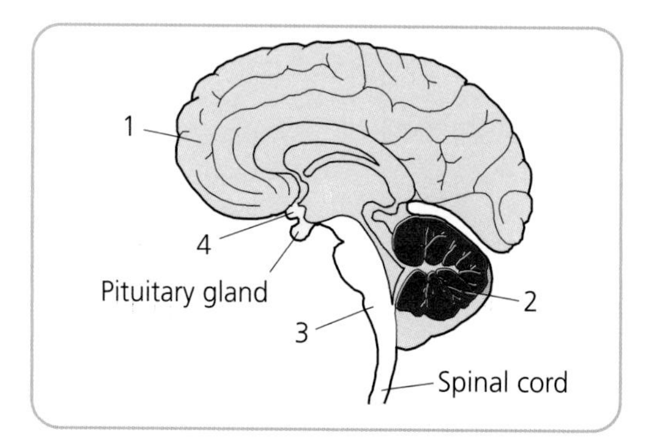

Figure 13.1

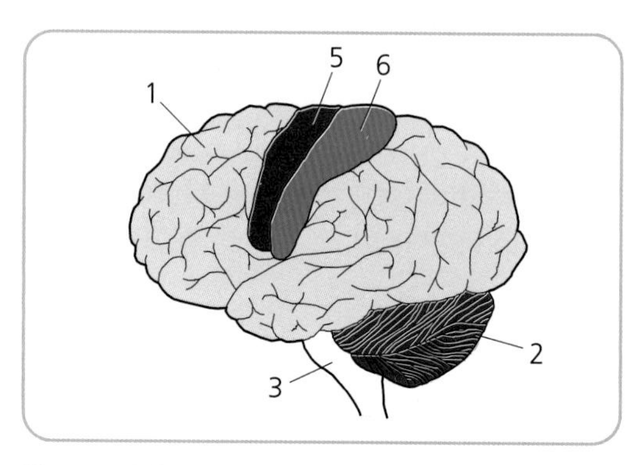

Figure 13.2

Use the key to identify:
1. the structures of the brain,
2. the function of each structure.

Key

1. **Cerebrum** – site of conscious responses and higher centres.
2. **Cerebellum** – centre of balance and co-ordination of movement.
3. **Medulla** – site of the vital centres for breathing and control of heart rate.
4. **Hypothalamus** – centre for regulation of water balance and temperature.
5. **Motor strip** – an area in the cerebrum that sends impulses to the muscles.
6. **Sensory strip** – an area in the cerebrum that receives impulses from the senses.

Structure and Function of the Nervous System

The brain, spinal cord and nerves

Figure 13.3 shows the brain, spinal cord and nerves.

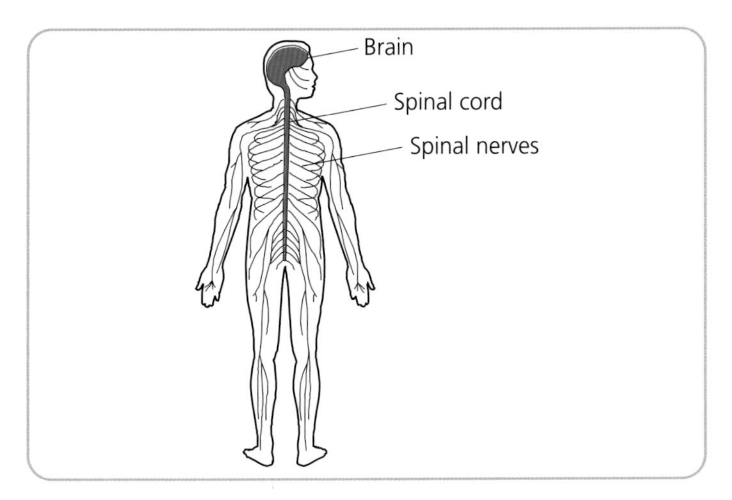

Figure 13.3

Key Points

- The brain and spinal cord form the **central nervous system** (CNS).
- **Nerves** are formed from bundles of nerve cells (neurones).
- A stimulus (touch, chemicals on the tongue, etc.) sets off an impulse in a neurone.
- Nerves carry information in the form of impulses from the senses to the CNS.
- The CNS sorts out the information that arrives as impulses from the senses.
- Nerves carry messages in the form of impulses from the CNS to the muscles.
- The muscles make the appropriate responses.
- These responses can be the co-ordinated movement of the arms, legs etc. or the contraction of muscles that control peristalsis.

Reflex action and the reflex arc

Reflex actions do not involve conscious thought. The neurones involved in reflex actions loop the impulses across the spinal cord and by-pass the brain.

Reflex actions include coughing, sneezing, iris response to light, knee-jerk, etc.

Reflex responses are rapid and their function is to protect the body from damage.

The neurones involved in reflex actions are shown in Figure 13.4 of the reflex arc.

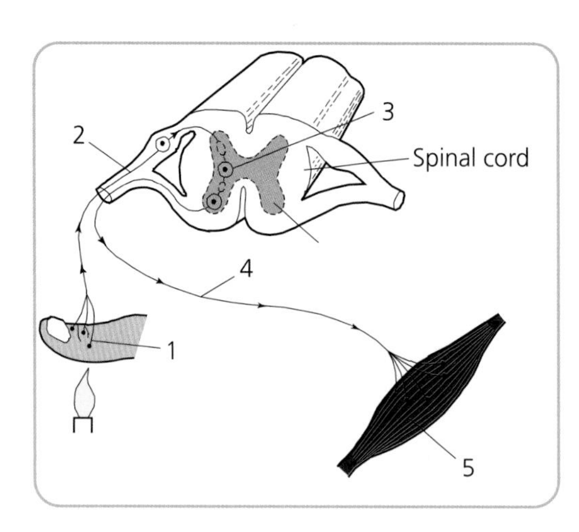

Figure 13.4

Use the key to identify: 1. the neural pathway involved, 2. the function of each structure.

Key

1. **Receptor** – sensory endings of a neurone detects a stimulus.
2. **Sensory neurone** – carries impulses to the spinal cord.
3. **Relay fibre** – carries impulses to a motor neurone.
4. **Motor neurone** – carries impulses to an effector.
5. **Effector** – brings about the response.

Effectors are muscles that contract and give rise to movements that save the body from damage.

Stop Think Learn

1 Copy the diagram of the brain, leaving out the key.

Copy and complete this table by naming each numbered structure and giving its function. The cerebrum has been completed for you.

Number	Structure	Function
1	Cerebrum	Site of conscious responses and higher centres (reasoning, caring, etc.)
2		

2 Copy and complete the blanks in the sentences.

The CNS is made up from the _________ and _____________________.

Information is carried by ______ to the CNS from the _____________.

The CNS _____________ the information.

Nerves carry ___________ from the CNS to the ___________.

3 What is the function of reflex responses?

4 Copy the diagram of the reflex arc, leaving out the key.

Copy and complete this table by naming each numbered structure and giving its function. The receptor has been completed for you.

Number	Structure	Function
1	Receptor	Detects a stimulus
2		

Temperature Regulation as a Negative Feedback Mechanism

Figure 13.5 outlines how temperature is regulated in the human body.

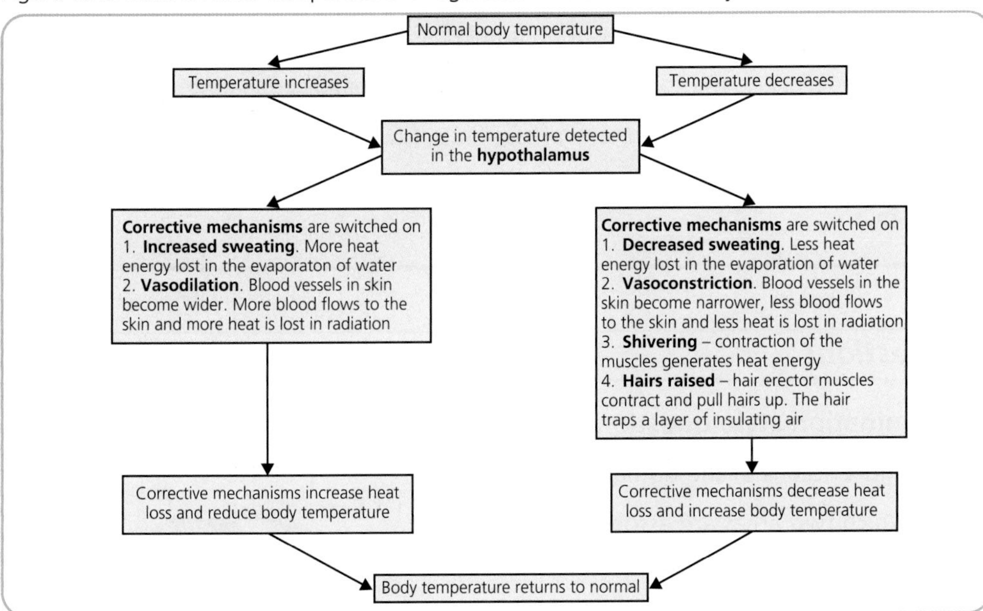

Figure 13.5

Temperature regulation is a negative feedback mechanism. Stages in a negative feedback mechanism include:

(1) change in temperature is detected in the hypothalamus,

(2) nerve pathways from the hypothalamus switch on the corrective mechanisms,

(3) corrective mechanisms return the temperature to normal,

(4) the return to normal temperature is detected in the hypothalamus, and

(5) the corrective mechanisms are switched off.

When the temperature is returned to normal the corrective mechanisms are switched off (negated).

Stop Think Learn

1 Where is change in body temperature detected?

2 Name two corrective mechanisms that help to reduce body temperature.

3 Explain how hairs being raised helps in temperature regulation.

4 Copy and complete the blanks in the sentences.

In temperature regulation when the ____________ returns to normal this is detected in the ____________________ and the corrective mechanisms are _____________.

This method of control is described as a _____________ feedback mechanism.

Hints and Tips

'Marks & Spencer' is a good way to remember Motor and Sensory strips in the correct order in the cerebrum, starting from the front of the brain.

In negative feedback, the corrective mechanisms are negated (stopped).

Make 'flash cards' for the words and phrases in **bold** in the text.

Motor is associated with movement and movement is associated with muscles. Think motor, think muscles.

Questions

Examination style questions

1 Figure 13.6 is of the human brain.

Which two areas are associated with co-ordination of movement?

A P and T

B Q and S

C T and S

D P and R

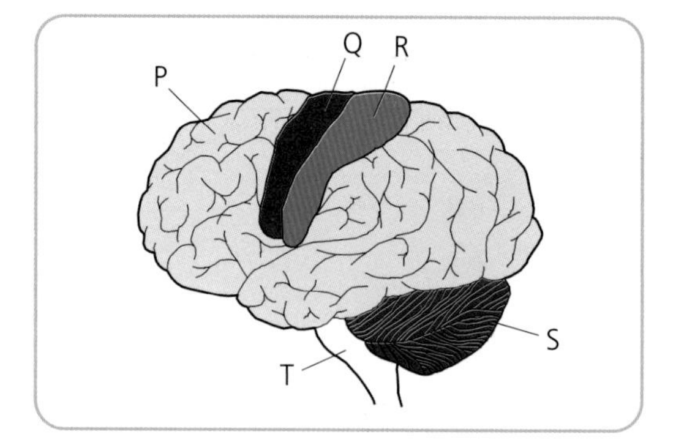

Figure 13.6

2 The following list refers to temperature regulation in a human.

List

1 constriction of blood vessels to the skin

2 dilation of blood vessels to the skin

3 increased sweating

4 decreased sweating

5 increased blood flow to the skin

6 decreased blood flow to the skin

Which of the above have a role in temperature regulation when body temperature is increased?

A 1, 3 and 5

B 1, 4 and 6

C 2, 3 and 5

D 2, 3 and 6

Questions continued

3 Figure 13.7 represents the control system involved in returning body temperature to normal after an increase.

a) State the exact location of the detection centre. (1)

b) Describe the change in blood flow to the skin with an increase in body temperature and name the mechanism that brings about this change. (1+1)

c) Describe the negative feedback mechanism that takes place after the body temperature returns to normal. (2)

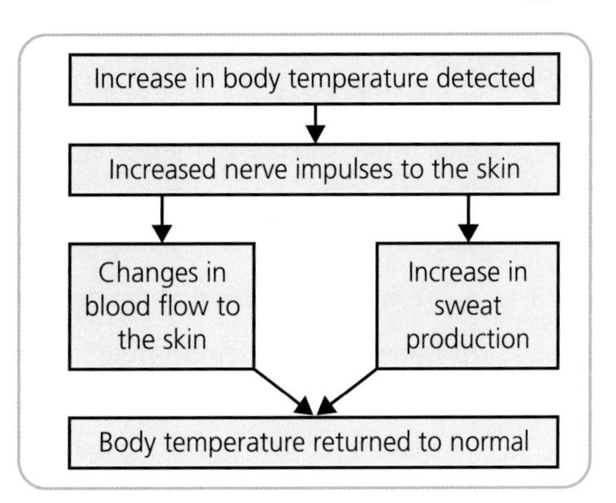

Figure 13.7

4 Figure 13.8 represents the structures involved in a reflex response. When food enters the alimentary canal it sets off reflex arcs that lead to the contraction of muscles. This causes food to be moved along the length of the intestines.

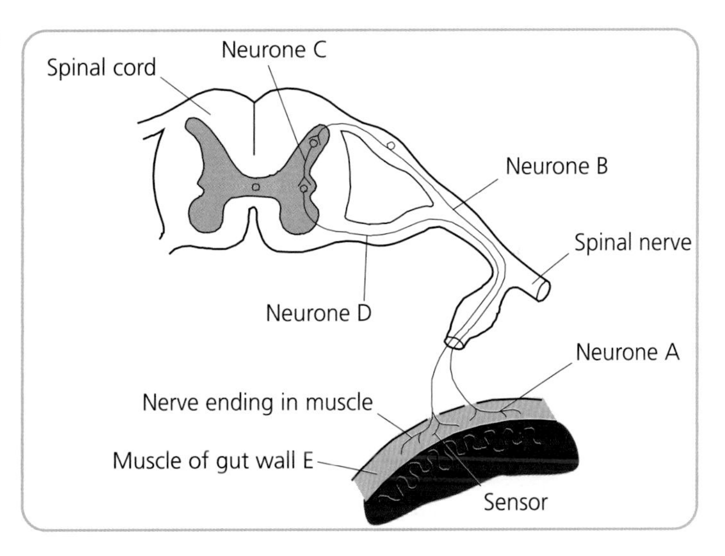

Figure 13.8

Complete this table which relates to the reflex response. *(2)*

Name of structure	Letter
Effector	
	C
Sensory neurone	
	D
	A

5 Describe the functions of the cerebrum, cerebellum and medulla of the brain. *(5)*

Answers

Answers to examination type questions with commentary

1 You should know that the cerebellum co-ordinates movement together with the motor area of the cerebrum. These areas are shown by Q and S. **Answer = B**

2 Temperature increase is controlled by dilation of blood vessels to the skin, increased sweating and increased blood flow to the skin. Numbers are 2, 3 and 5. **Answer = C**

3 a) You should know that temperature is detected by the hypothalamus.

b) An increase in temperature means that the body has to cool down and this involves loss of heat energy. To lose more heat more blood has to flow to the skin.

1st mark is for saying 'More blood flows to the skin'. 2nd mark is for knowing that this is caused by the 'Blood vessels widening' or using the term 'Vasodilation'.

c) You should know that negative feedback means that after return to normal has been detected then corrective mechanisms are negated (stopped).

The return to normal is detected by the hypothalamus 1st mark.

And the corrective mechanisms are switched off 2nd mark.

4 You have to be able to identify the structures involved in the reflex arc.

Receptor is A; sensory neurone is B; relay fibre is C; motor neurone is D; effector is the muscle E. 5 correct = 2 marks; 4 or 3 correct = 1 mark.

5 Describe the functions of the cerebrum, cerebellum and medulla in the brain.

In extended writing questions you must identify all the parts that make up the question. These are:

- Describe the functions of the cerebrum.
- Describe the functions of the cerebellum.
- Describe the functions of the medulla.

A1 Cerebrum is the site of conscious responses.

A2 Cerebrum is the site of higher centres (reasoning, etc.).

A3 Discrete area of the cerebrum relates to motor function / muscle function.

A4 Discrete area of the cerebrum relates to sensory information.

Maximum of 3 marks.

B1 Cerebellum is the centre of balance.

B2 Cerebellum is the centre for co-ordination of movement.

Maximum of 1 mark.

C1 Medulla is the centre for breathing.

C2 Medulla is the centre for heart rate.

Maximum of 1 mark. Maximum total = 5 marks.

Chapter 14

SKILLS IN KNOWLEDGE AND UNDERSTANDING IN MULTIPLE CHOICE QUESTIONS

In the National Examination, Section A consists of multiple choice questions.

The Course Specifications for Section A are as follows:

Section A of the Intermediate 2 Biology examination consists of 25 multiple choice questions. 9–11 are Problem Solving / Practical Abilities the rest being Knowledge and Understanding (KU). This part of the exam is answered on a separate Marking Grid.

Your examination script will be personalised. Details of your name, candidate number, centre number, and so on, will already be printed out onto the Marking Grid.

Part of an Answer sheet is shown in Figure 14.1.

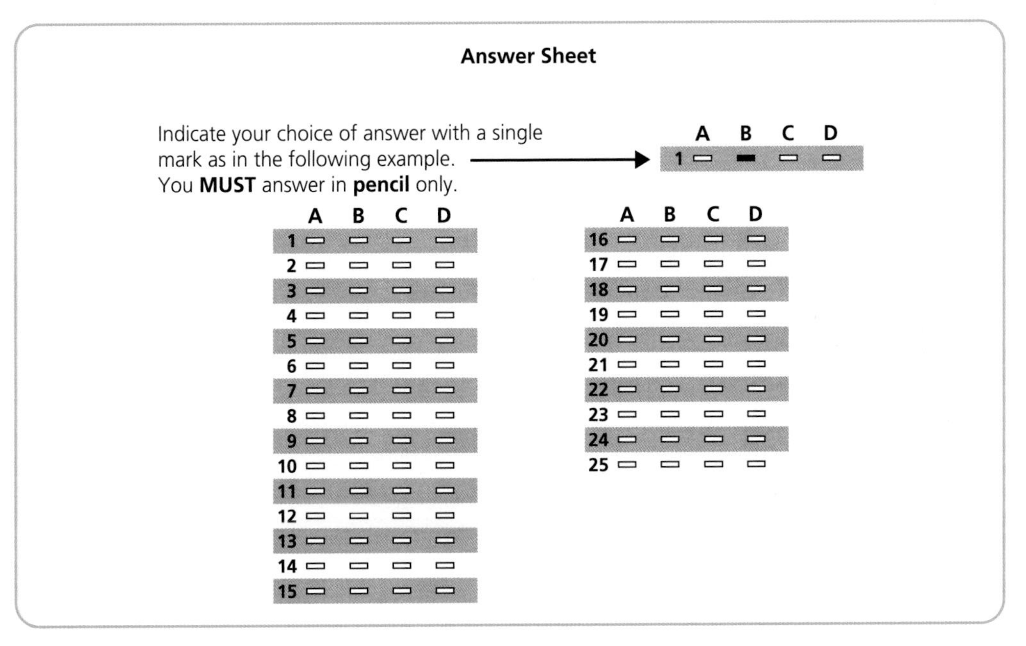

Figure 14.1

Sample question

Which part of the brain is involved in the control of heart rate?

A	Cerebellum
B	Medulla
C	Hypothalamus
D	Cerebrum

The correct answer is **B – Medulla**.

The answer **B** has been clearly marked in Figure 14.1 with a horizontal **BLACK** line.

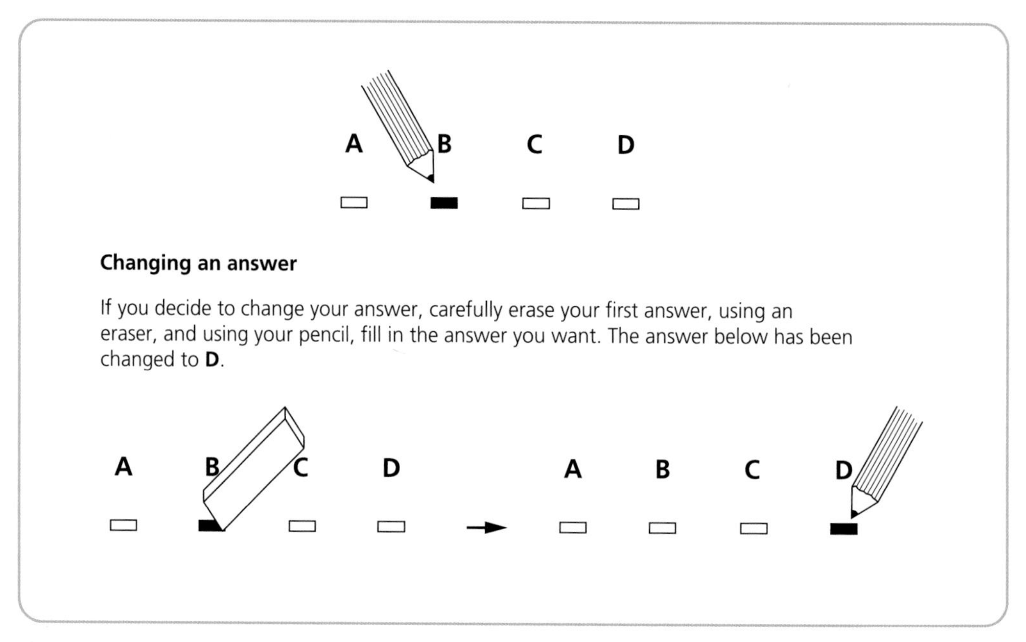

Figure 14.2

Strategy for multiple choice knowledge and understanding questions

Key Points

- Always read each question very carefully. Highlight/underline key words.
- The sequence of questions usually follows the order in which topics appear in the Arrangements Documents. (This outlines the requirements of the Intermediate 2 Biology course and assessment.)
- A good strategy for KU questions is to see if you can answer the question **without looking** at the four choices.
- If you think that you know the answer, then look for the option that best matches your answer.
- If a question is giving you difficulty, try first to remove any of the options that you think are clearly wrong. Continue to try to remove options until you are left with only one. This has to be your answer, even if you do not know for certain that it is correct.
- There is usually a fairly equal distribution of answers A, B, C and D. If you have time at the end of the examination it is worth checking this. Examples of multiple choice questions together with a commentary are given in the text that follows.

Example

Example 1

Which line in the table below correctly describes the part of the body where excess proteins are broken down and what they are broken down into?

	Location	Excess proteins broken down into
A	liver	urea
B	kidney	urea
C	liver	amino acids
D	kidney	amino acids

You should know proteins are made up of amino acids and that excess amino acids are deaminated in the liver. Urea is produced as a result of deamination.

You are looking for liver and urea **Answer = A**

Example 2

Which of the following conditions in a greenhouse would produce earlier crops?

You should know that conditions that would increase growth would have to increase photosynthesis. These include supplementary lighting, additional carbon dioxide and increase in temperature (heating). You are looking for any one of these three:

A glass shading

B cool air conditioners

C additional oxygen

D additional carbon dioxide **Answer = D**

Example 3

Figure 14.3 represents the transmission of sex determining chromosomes from parents to offspring.

Which line in the following table correctly identifies the sex chromosomes for the gametes P, Q, R and S?

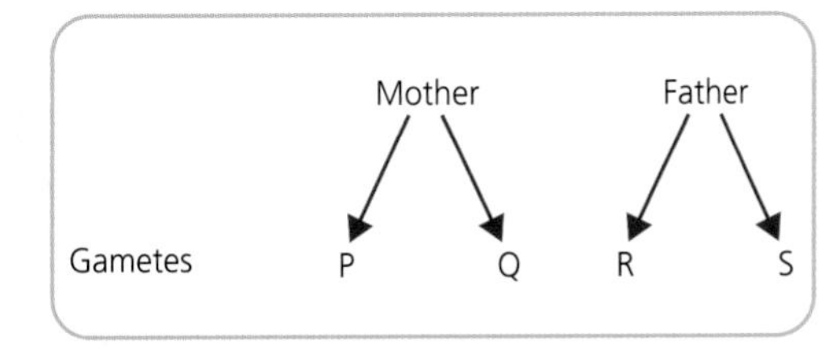

Figure 14.3

Example continued ➢

Example continued

	Gamete P	Gamete Q	Gamete R	Gamete S
A	XX	XX	XY	XY
B	X	X	Y	X
C	X	Y	X	Y
D	X	Y	X	X

You should know the genotype of the mother is XX and that she forms gametes that contain one of her X chromosomes. The genotype of the father is XY and he can form gametes either with an X or with a Y chromosome.

You are looking for one of the following:

X X X Y OR X X Y X

Answer = B

Example 4

The following stages occur during photosynthesis.

W glucose is formed

X water is broken down to produce hydrogen

Y glucose is converted to starch

Z hydrogen is combined with carbon dioxide

What is the correct order of the stages?

A W Z X Y

B Z Y X W

C X Z W Y

D Y X Z W

Now is the time to go through your knowledge on photosynthesis.

Light energy absorbed by chloroplasts; hydrogen produced from splitting of water; ATP produced; hydrogen combines with carbon dioxide; energy of ATP required; glucose is formed; glucose stored as starch.

Not all of these stages are included in the choices given. However, from those given you are looking for the sequence X Z W Y. **Answer = C**

Example continued ➢

Example continued

5 Figure 14.4 shows some structures in a villus.

Which line in the table correctly identifies the products of digestion that pass into structures X and Y?

You should know that X is a lacteal and Y is a blood capillary.

Products of fat digestion (glycerol and fatty acids) are absorbed into the lacteals.

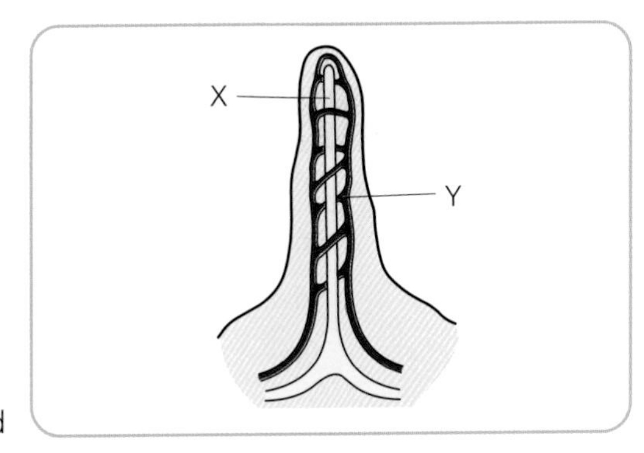

Figure 14.4

Products of starch and protein digestion (glucose and amino acids) are absorbed into the capillaries.

You are looking for any one of these products of digestion being absorbed into the correct structure.

	X	Y
A	glucose	amino acids
B	glycerol	fatty acids
C	amino acids	glycerol
D	fatty acids	glucose

Answer = D

6 Figure 14.5 shows a site of gas exchange in the lungs.

Which line in the following table shows the correct relative concentrations of oxygen at W, X and Y?

You should know that W shows the air entering the alveolus and the oxygen concentration will be relatively high.

Arrows in the capillary show blood is flowing from X to Y.

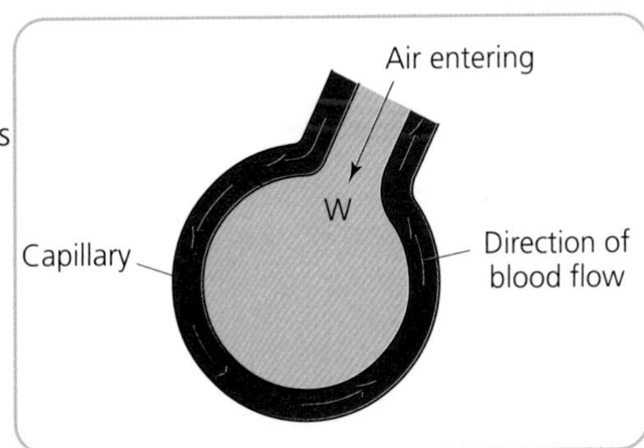

Figure 14.5

Blood returning to the lungs at X will have given up its oxygen and concentration will be relatively low. As it flows through the capillary, blood will take up oxygen and by the time it reaches Y the concentration will be relatively high.

Example continued ➢

Example continued

Look for the answer that best matches the information.

	Relative concentration of oxygen at W	at X	at Y
A	high	low	high
B	low	high	low
C	low	high	high
D	high	high	low

Answer = A

Chapter 15

SKILLS IN KNOWLEDGE AND UNDERSTANDING IN STRUCTURED QUESTIONS

In the National Examination, Section B consists of structured questions. The Course Specifications for Section B are as follows:

This section will contain structured questions with an allocation of 65 marks. Between 45–50 marks testing Knowledge & Understanding (KU) and between 15 and 20 marks testing problem solving and practical abilities

Problem Solving/Data Handling questions should involve one source of data.

Practical Abilities/Experimental situation questions should be set in less familiar contexts.

(Skills in Data Handling and Experimental situations are outlined in Chapters 17 and 18.)

Structured Knowledge and Understanding Questions

Questions that begin with NAME, STATE, GIVE, IDENTIFY, WHAT, WHICH etc. are usually of a lower level of demand (easier questions).

Questions that begin with DESCRIBE, EXPLAIN (WHY/HOW), are usually of a higher level of demand (harder questions). To gain a pass at Grade A or Grade B, you must answer questions of this type correctly.

An indication of the number of points required for an answer can be determined by:

1 The number of marks awarded for a question.

2 The number of lines provided in the answer space.

Hints and Tips

Always read the questions carefully. Highlight/underline key words.

Answer the question that is asked, not the question that you think is asked.

Try to write down all the facts that you consider relevant to the question.

Marks are not deducted in Section B for additional information, unless it contradicts the correct answer.

Abbreviations are acceptable for many terms – DNA for deoxyribonucleic acid, ADH for antidiuretic hormone. These are shown in the Arrangements Document.

A wide variety of question types are used and examples of these are included within this chapter.

Example

Example 1

These are questions that are straightforward recall. These you **must** study for.

'Flash cards' are very useful for recall.

(a) Muscle fatigue may occur during exercise. Name the chemical that results in muscle fatigue.

______________________ *(1)*

(b) Name the structures in the lungs where gas exchange takes place.

______________________ *(1)*

(c) Name the type of neurone that links receptors in the sense organs to the CNS. *(1)*

____

Answers: (a) Lactic acid, (b) alveoli, (c) sensory neurone.

Example 2

The following word equation shows the first stage of photosynthesis.

$$\text{water} + \text{ADP} + \text{Pi} \xrightarrow{\text{light energy}} \text{oxygen} + \text{hydrogen} + \text{ATP}$$

(a) Name this stage of photosynthesis.

______________________ *(1)*

(b) Describe what happens to each of the products.

Oxygen ______________________

______________________ *(1)*

Hydrogen ______________________

______________________ *(1)*

ATP ______________________

______________________ *(1)*

Commentary

(a) You have to know that the splitting of water is called photolysis.

(b) These questions use **Describe** and are of a higher level of demand.

You will have to show clearly that you know what **happens** to each end product.

The fact that two lines are left for each answer also suggests this.

Oxygen is formed as a by-product and is released into the air **or** is used in respiration.

Hydrogen is picked up by a hydrogen carrier and is combined with carbon dioxide.

ATP is broken down to form ADP and Pi and the energy released is used for carbon fixation **or** glucose formation.

Example *continued* ➢

***Example** continued*

Example 3

Explain why diffusion is important to a leaf mesophyll cell in the light.

____

____

____ *(2)*

Commentary

Note that the question has three lines for the answer and a 2 mark allocation.

An answer such as '*Diffusion allows the movement of substances into or out of the cell*'. does not answer the question. This simply defines diffusion and fails to give an explanation.

'Explain why' questions have to be answered in two parts. They can be treated as questions that require an **IT IS BECAUSE** followed by an **AND**.

The question refers to leaf cells and to light. The answer must refer to photosynthesis and the substance that is diffusing into the cell must be required for photosynthesis.

The substance must be carbon dioxide.

Answer

IT IS BECAUSE carbon dioxide will enter the cell by diffusion. *(1st mark)*

AND carbon dioxide is required to make glucose in photosynthesis. *(2nd mark)*

Example 4

To produce early crops in horticulture the air in greenhouses is enriched with carbon dioxide. Explain how this would make the crop grow faster.

____

____ *(2)*

Commentary

Note this question has two lines and a 2 mark allocation.

'Explain how' questions should be treated in the same way as 'Explain why' questions.

The stem of the question tells you that carbon dioxide has been increased. What you have to do is connect this to faster growth. The answer must refer to photosynthesis.

A further tip is that you have to use the word '**more**' in your answer. This is needed to explain the faster growth of the crops.

***Example** continued ➢*

Example continued

Answer

IT IS BECAUSE with more carbon dioxide photosynthesis will increase AND there will be more food for growth.

Questions 3 and 4 are Grade A questions. These are the questions that have to be answered in order to gain a pass at the highest grade.

Other styles of question

1 *Complete a table*

Bony fish can live in freshwater or marine habitats.

The list contains methods used by bony fish to overcome osmotic problems.

List

1 Secreting excess salts.
2 Excreting copious urine.
3 Producing very dilute urine.
4 Drinking surrounding water.

Use the information in the list to complete this table.

Habitat	Osmotic problem	Methods used by bony fish to overcome osmotic problems
	Water gain by osmosis	
	Water loss by osmosis	

(2)

Commentary

You have to know that if the fish is losing water it is in sea water and if gaining water it is in freshwater. This allows habitats to be identified. *(1st mark)*

Now from the list you know that 2 and 3 relate to freshwater fish and 1 and 4 to sea water fish. All 4 correct = 2 marks, 3/2 correct = 1 mark.

2 *Underline options to make a sentence correct.*

In the lungs haemoglobin [combines with / releases] oxygen at [high / low] oxygen levels. *(1)*

Commentary

This is in the lungs. You should know that haemoglobin combines with oxygen in the lungs because oxygen levels are high in the lungs.

Answer

Underline combines with and high

Example continued ➢

***Example** continued*

3 *True or False with correction.*

Decide if each of the following statements about temperature regulation in the body is **TRUE** or **FALSE** and tick (✓) the appropriate box in the table.

If you decide the statement is **FALSE**, write the correct word in the **Correction** box to replace the word underlined in the statement.

Statements	True	False	Correction
External temperature is detected by receptors in the skin			
The area of the brain which regulates body temperature is the medulla			
Blood vessels in the skin constrict in response to an increase in external temperature.			

(3)

Commentary

First statement is **TRUE**. Tick box.

Second statement is incorrect. Tick **FALSE** and insert 'hypothalamus' in correction.

Third statement is incorrect. Tick **FALSE** and insert 'dilate' in correction.

One mark for each correct answer.

Chapter 16

SKILLS FOR EXTENDED WRITING

In the National Examination, Section C consists of extended response questions. The Course Specifications for Section B are as follows:

This section will consist of four extended response questions to test the candidates' ability to select, organise and present relevant knowledge.

Candidates have to answer two extended writing questions with a choice of one from two within each question. Section C will have an allocation of 10 marks (5 marks for each extended response question).

Extended Response Questions

A typical choice in the first of the extended response questions is shown in Figure 16.1.

SECTION C

Both questions in this section should be attempted.

Note that each question contains a choice.

Questions 1 and 2 should be attempted on the blank pages which follow.
Supplementary sheets, if required, may be obtained from the invigilator.

1. Answer **either** A **or** B.

   A. The diagram below shows a section through the human heart.

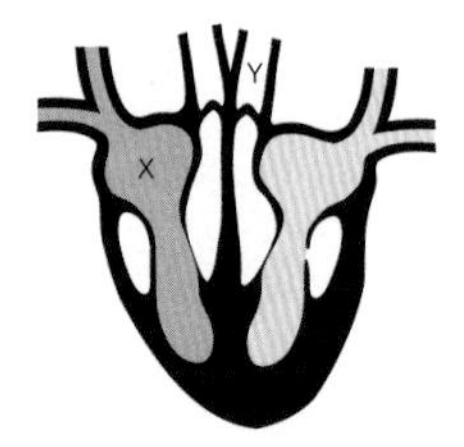

Describe the pathway of blood through the heart and associated structures starting at X and finishing at Y. There is no need to mention the valves. **5**

OR

B. Urine production occurs in the kidney. The diagram below shows the structures of a nephron and its blood supply.

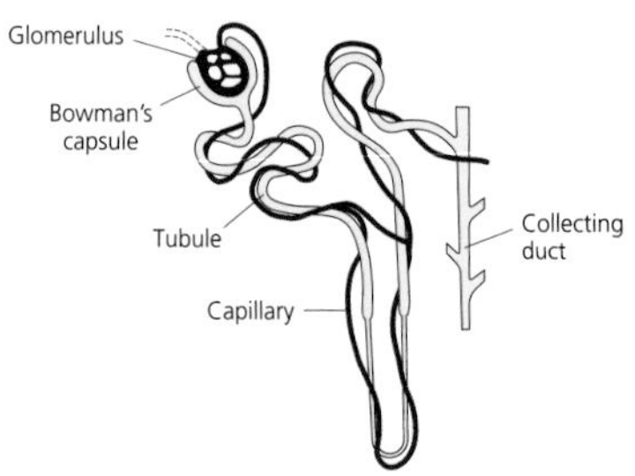

Describe how the nephron produces urine. There is no need to mention the role of ADH. **5**

Figure 16.1

In this first extended response question, 'the examiner gives support to the candidate in the form of a labelled diagram', and so on to help stimulate the candidate's memory.

The candidate has to choose one out of the two questions offered.

Candidates in their response **must address all areas** within the question in order to gain full marks. The procedure outlined will help in answering extended response questions.

Hints and Tips

Identify clearly all the areas that you must write about.

Use the words as used in the question – **name**, **describe**.

Make a **list** of the areas.

As you answer the question, **tick off each area** as it is covered.

Examples

Example 1A (Figure 16.1)

Areas which must be addressed:

1. **Describe** the pathway through the heart starting at X and finishing at Y.
2. **Describe** the pathway through the associated structures.

Example 1B (Figure 16.1)

In this example, Figure 16.1 can be used to allow identification of the areas to be addressed.

Areas that must be addressed:

1. **Describe** what happens at the glomerulus and Bowman's capsule.
2. **Describe** what happens in the tubule.
3. **Describe** the role of the capillary.
4. **Describe** what happens in the collecting duct.

Stop Think Learn

1. Identify all the areas that you must write about in Figure 16.2. The diagram outlines photosynthesis.
   Name and describe Stage 1 and Stage 2.
2. A typical choice in the second of the extended response questions is shown on the following page. Answer either A or B.

   You may include labelled diagrams where appropriate.

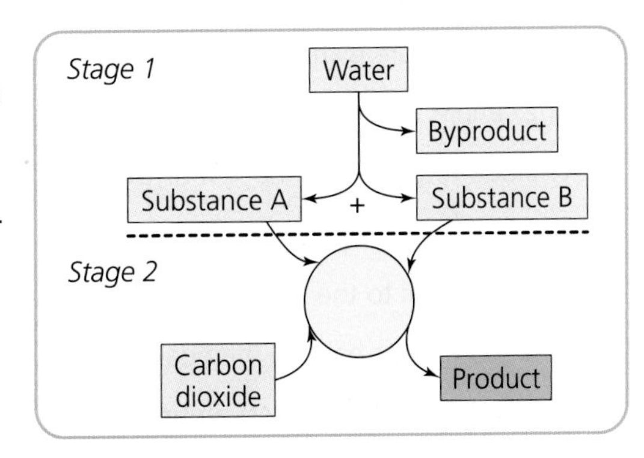

Figure 16.2

A. Describe the structures of arteries, veins and capillaries. Give the function of each of these types of blood vessel. *(5)*

OR

B. Freshwater bony fish have a water balance problem. State the water balance problem and describe how these fish overcome the problem. *(5)*

There are no diagrams given with these questions. However, the method of answering them is the same as in the first type of question.

Examples

Example 2A

Areas which must be addressed:

Describe the structure of an artery.

Describe the structure of a vein.

Describe the structure of capillaries.

Give the function of an artery.

Give the function of a vein.

Give the function of capillaries.

Example 2B

Areas which must be addressed:

State the water balance problem of freshwater bony fish.

Describe how these fish overcome the problem.

Stop Think Learn

Identify all the areas that you must write about in the examples below.

1 Describe adaptations to habitat and niche as illustrated by Darwin's finches and desert plants.

2 Describe the commercial and industrial uses of cells.

Cut-off scores

Within each area that you have to write about there is a **cut-off score**.

The cut-off score is the maximum mark that can be gained within that area.

For example, for question 1A in figure 16.1.

A1 Blood returns from the rest of the body to X, the right atrium.

A2 Blood goes to the right ventricle.

A3 Blood goes to the lungs via the pulmonary artery.

A4 Blood returns back to the heart.

A5 Blood returns to the left atrium.

Maximum 3 marks.

B1 Blood enters the left atrium.

B2 Blood goes to the left ventricle.

B3 Blood goes to vessel Y which is the aorta.

Maximum 2 marks.

Maximum total = 5 marks.

Notice, by the use of cut-off scores, if you failed to give information on flow through the left side then your maximum mark could only have been 3 marks.

The Marking Instructions for question 2B (page 152) on water balance in fish are shown below.

A1 Tissues of fish are hypertonic/at a lower water concentration to the surroundings.

A2 Water moves down the water concentration gradient.

A3 Water enters the fish continuously.

A4 Water moves by osmosis.

A5 Cells may swell and rupture or osmotic balance upset.

Maximum 3 marks.

B1 Fish are able to osmoregulate.

B2 Large volumes of urine are produced.

B3 Urine is very dilute.

Maximum 2 marks.

Maximum total = 5 marks.

Notice, by the use of cut-off scores, if you failed to give information on how fish overcome the problem then your maximum mark could only have been 3 marks.

Hints and Tips

In order to gain the maximum 5 marks in extended response questions the following must apply:

Written notes must address **all areas** identified within a question.

Include all the information that you think is relevant. This increases the chance of achieving the **cut-off scores** within each area. Marks are not deducted in Section C for additional information, unless it contradicts the correct answer.

Chapter 17

PROBLEM SOLVING

The National Course specifications for Intermediate 2 describe Problem Solving as follows:

Questions relating to each of the following points will be included in the course examination in order to test the candidate's ability to:

1. *Select relevant information from texts, tables, charts, graphs and/or diagrams.*
2. *Present information appropriately in a variety of forms, including written summaries, extended writing, tables and/or graphs.*
3. *Process information accurately using calculations where appropriate. Calculations to include percentages, averages, and/or ratios. Significant figures and units should be used appropriately.*

Course details state that in Section A of the examination 9–11 questions will test Problem Solving (PS) and Practical Abilities (PA) and in Section B between 15–20 marks will test PS and PA. This means that around 25% to 30% of the examination tests PS and PA. These are skills you have to develop. The examples that follow, together with a commentary, give support to answering questions set to test each of the points listed above.

Selecting Information

Selecting information from a table

Example

The tables show water loss and gain by a desert rat over a 24-hour period.

Water loss	Mass of water lost (g)
Urine	12
Faeces	3
Exhalation	45

Water gain	Mass of water gain (g)
Food	6
Metabolic water	54

What evidence supports the statement that the desert rat is in water balance over the 24-hour period? *(1)*

You should know that to be in water balance the water loss must equal the water gain.

The loss adds up to 60 g and the gain to 60 g.

Answer = The evidence is that water loss and water gain are equal at 60 g.

Selecting information from a pie chart

A pie chart is given either as a percentage or in degrees.

The whole pie is 100% or 360°. Three-quarters is 75% or 270°. A half is 50% or 180°.

A quarter is 25% or 90°. An eighth is $12^1/_2$% or 45°.

Examples

Figure 17.1 shows the contents of a type of yoghurt.

What are the percentages of water and sugar in the yoghurt?

Step 1 – Identify water from the key.

Step 2 – Water is three-quarters of the pie = 75%.

Step 3 – Identify sugar from the key.

Step 4 – Sugar is an eighth = $12^1/_2$%.

Answer = Water = 75% and sugar = $12^1/_2$%

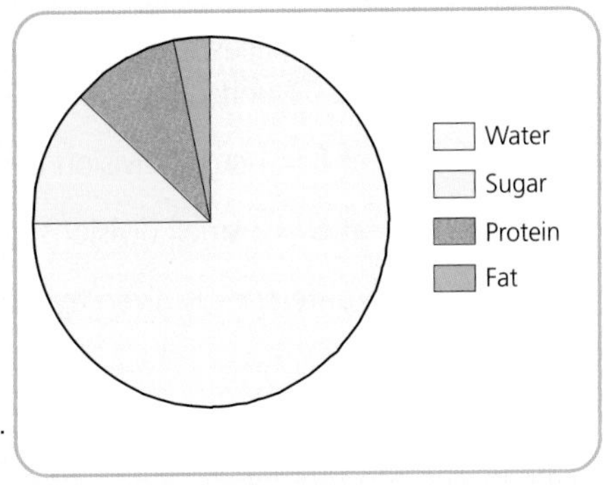

Figure 17.1

Selecting information from a graph

Example

Example 1

Figure 17.2 shows the effects of increasing carbon dioxide concentration on the rate of photosynthesis at different light intensities and temperatures.

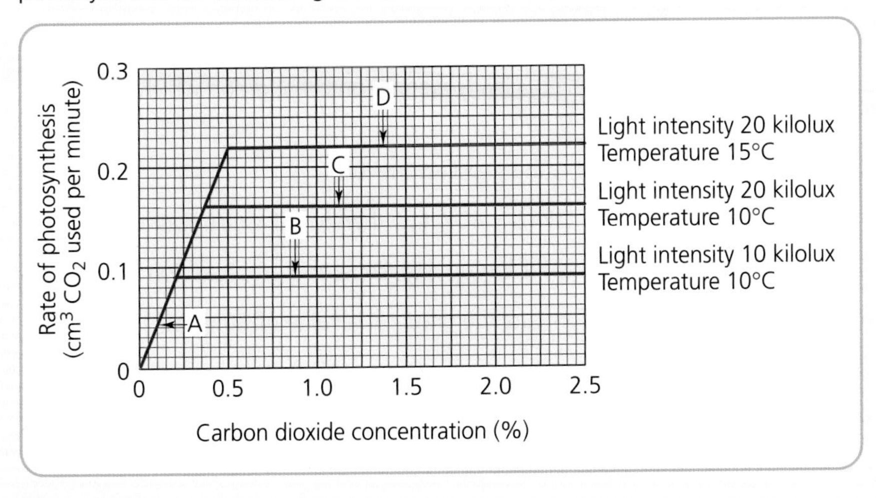

Figure 17.2

Example continued ➢

***Example** continued*

From the graph, calculate the difference in the rates of photosynthesis between points B and D.

For all graphs, first, find the value of the smallest division on the axis required.

Step 1 – Value of smallest division on the y-axis: 10 small divisions = 0.1 units, therefore 1 small division = 0.01 units.

Step 2 – Rate at B = 1 small division less than 0.1 = 0.1 – 0.01 = 0.09.

Step 3 – Rate at D = 2 small divisions greater than 0.2 = 0.2 + 0.02 = 0.22.

Answer = The difference is given by 0.22 – 0.09 = 0.13 cm^3 CO_2 used per minute.

Example 2

An investigation was carried out into the uptake of sodium ions by animal cells.

Figure 17.3 shows the rates of sodium ion uptake and breakdown of glucose at different concentrations of oxygen.

Calculate the number of units of sodium ions that are taken up over a 5-minute period when the concentration of oxygen in solution is 2.0%.

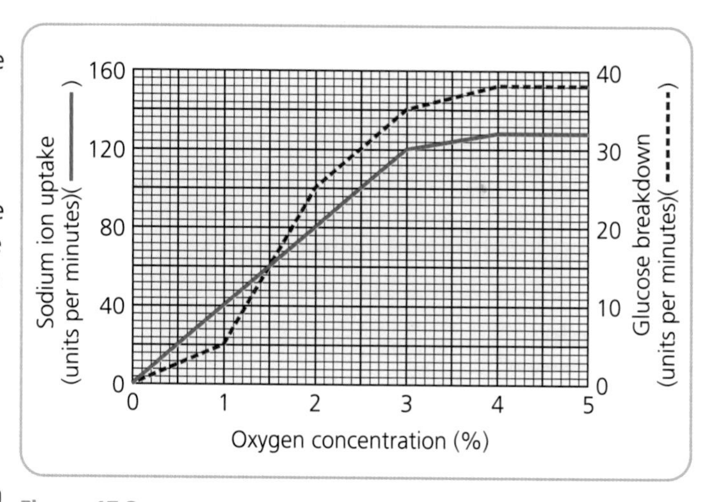

Figure 17.3

If two graphs and two y-axes are labelled, clearly identify onto the graph which graph is which and which axis refers to which graph.

Step 1 – Scale for sodium ion uptake is the y-axis to the left.

Value of smallest division: 10 small divisions = 40 units, therefore 1 small division = 4 units.

Step 2 – The solid line in Figure 17.3 is for ion uptake. Label this graph as 'ion uptake'.

Step 3 – The 2% oxygen concentration cuts the graph at a value of 80 units.

Step 4 – This is the value for 1 minute. You were asked to calculate for 5 minutes.

Answer = 80 × 5 = 400 units.

Select information from a diagram

Example

Figure 17.4 represents part of a foodweb in an ecosystem.

This question requires the knowledge that arrows in a foodweb show the direction of energy transfer.

(a) Identify an omnivore from the foodweb. *(1 mark)*

You must know that an omnivore gets its energy by eating both plants and animals.

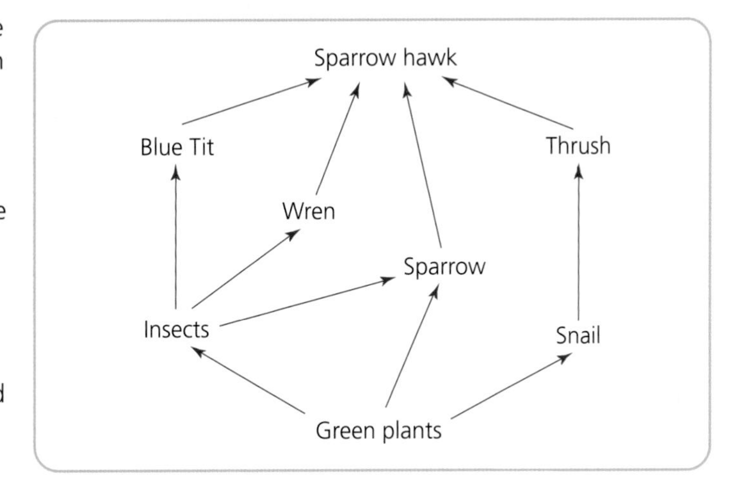

Figure 17.4

Step 1 – Start at the plants, it can only be the snail, insects or the sparrow.

Step 2 – Look at other connections from the sparrow, insects and snail.

The sparrow also gets energy by eating insects. **Answer** = Sparrow

(b) Explain how a decline in the sparrow hawk population leads to a decrease in the insect population. *(1 mark)*

Step 1 – Identify the energy sources of the sparrow hawk – blue tit, wren, sparrow and thrush.

Step 2 – Check to see if any of these get energy from insects – blue tit, wren, sparrow.

Answer = Fewer blue tits, wrens and sparrows are eaten and with an increase in their numbers more insects will be eaten.

Select information from a key

You should know that in this type of key you have to select from paired statements (refer to Figure 17.5).

Statements are numbered and as you follow the path down the key from number to number you eventually come to the correct answer.

You should also know how to work backwards from the key by following the numbers in reverse order.

Example

Figure 17.5 shows the leaves of some Scottish plants.

Identify any characteristics that are shown to be common to both sycamore and beech leaves.

Step 1 – Identify sycamore and work back to the start of the key noting characteristics: serrated edge – lobed – simple.

Step 2 – Repeat process for beech. Length less than double breadth – not lobed – simple.

1. Simple	(2)
Compound	(5)
2. Lobed	(3)
Not lobed	(4)
3. Smooth edge	ivy
Serrated edge	sycamore
4. Length about double the breadth	privet
Length less than double the breadth	beech

Figure 17.5

Answer = Only characteristic shown to be common is that leaves are simple.

Presenting Information

Presenting information as a graph

Graphs used in the National Examinations show slight differences from year to year, but the basic rules that apply are the same. The following procedures refer to line graphs. The same procedures apply to a bar graph.

The factor that was **controlled** in the experiment becomes the x-axis.

Each axis (x and y) is labelled by inserting **all the details listed** in the headings in the table of results.

If both the x and y-axes have values that begin at 0 then you must insert a 0 at both axis.

The scale selected must use at least half of the area available in the graph paper provided (loss of 1 mark).

The scale used must be of equal value: 0 to 5, 5 to 10, 10 to 15, etc.

If the values given do not begin at 0 then you do not have to make up a scale that starts at 0. For example, if a range of values is given from 450 to 650 units then the starting point on the axis is 450 and not 0.

The examiner helps the candidate in the selection of a suitable scale because the number of squares available within the graph paper relates to the range of values given.

All points must be plotted accurately. A dot on the exact spot or a thin X (not an amorphous blob).

All points must be joined. **Use a ruler**. Do not have thick lines or wiggly lines.

A graph to show line of best fit is not appropriate at Intermediate 2 level.

Only connect the line of the graph to the origin if this value has been given.

If you make a mistake with the graph do not panic or apply paper fluid. There is a second graph in the end pages to allow for a second attempt.

Example

The results show the effect of temperature on the rate of respiration by yeast.

The rate of respiration was measured by the volume of gas given off in 1 hour.

Temperature (°C)	10	20	30	40	50
Volume of gas collected after 1 hour (cm^3)	10	20	40	50	5

Using the results in the table, plot a line graph of volume of gas collected after 1 hour against temperature.

The factor controlled is temperature. This becomes the x-axis.

Label the x-axis by inserting all the details listed in the heading of the table of results: Temperature (°C).

To select a scale:

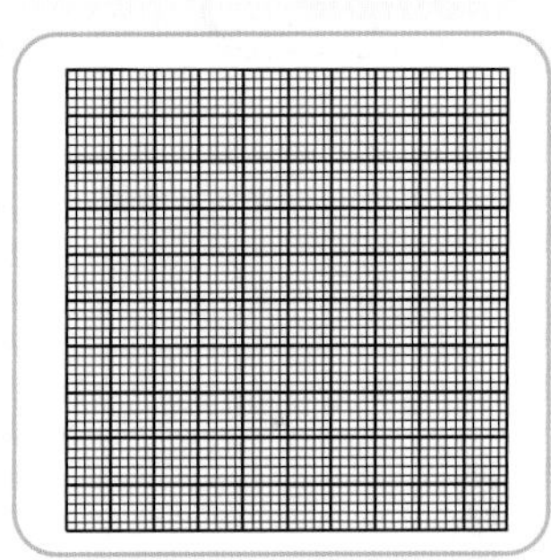

Count the number of squares available = 10.

Find the highest value used = 50°C.

Divide the value by the number of squares 50/10 = 5°C.

Each square has a value of 5°C if the graph starts at zero.

Label the y-axis by inserting all the details listed in the other heading in the table: Volume of gas collected after 1 hour (cm^3).

To select a scale:

Count the number of squares available = 10. Find the highest value used = 50.

Divide value by number of squares 50/10 = 5. Each square has a value of 5 units if the graph starts at zero.

Insert the scale: start from zero and increase by units of 5. Thus 0, 5, 10, to 50 units.

Plot the graph: use either a dot on the spot or a small X that bisects the spot.

Join the points using a ruler.

Processing Information

Averages

The average of a set of values is the value which best represents the situation. For example, you travel a distance of 20 miles by bus and it takes 1 hour. The bus does not travel at the same speed all the time. The bus stops at bus stops, at traffic lights, etc. yet the total time for the 20 miles journey was 1 hour. This is the average speed for the journey – 20 miles per hour.

Method for calculating averages

Step 1 – Calculate the sum/total of the set of values. = TOTAL

Step 2 – Calculate the number of values. = NUMBER

Step 3 – Calculate the average by dividing the TOTAL by the NUMBER of values.

$$\text{AVERAGE} = \frac{\text{TOTAL}}{\text{NUMBER}}$$

Answer: Average _______ UNITS

Example

Example 1

Find the average length of an earthworm from this table.

Sample	1	2	3	4	5	6	7	8	9	10
Length (mm)	10	11	13	25	26	11	12	15	15	12

Step 1 TOTAL of set of values = 130 mm

Step 2 NUMBER of values = 10

Step 3 $\text{AVERAGE} = \frac{\text{TOTAL}}{\text{NUMBER}} = \frac{130}{10} = 13$

Answer = Average = 13 mm.

If units are not shown in the answer then you MUST insert the correct units.

Example 2

Figure 17.6 shows cells under a microscope.

Calculate the average length of the cells.

In this type of example the TOTAL is the length of the field of view.

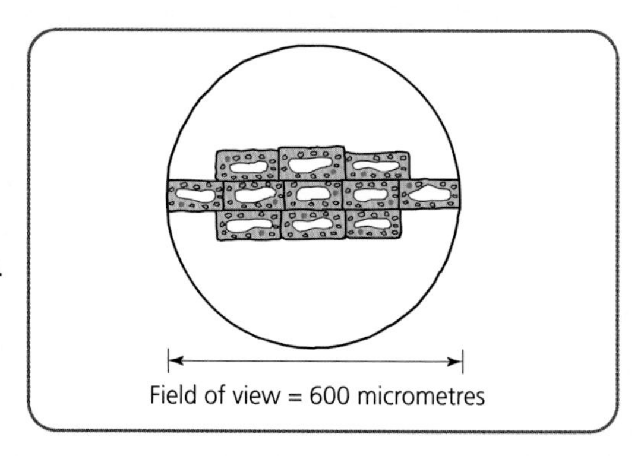

Figure 17.6

Example continued ➢

Example continued

Step 1 – TOTAL length of field of view = 600 micrometres

Step 2 – NUMBER of cells in field of view = 5

Step 3 – AVERAGE = $\frac{\text{TOTAL}}{\text{NUMBER}}$ $= \frac{600}{5} = 120$

Answer = Average = 120 micrometres.

Percentages

Percentages are another way of representing fractions and are based on hundredths.

'Per cent' means per hundred or 'for every hundred'.

Percentages represent **relative** and not absolute quantities.

Method for calculating percentages

Create a fraction first and then convert the fraction to a percentage.

To convert a fraction to a percentage, **multiply by 100%**.

Fraction $\frac{\text{Upper number}}{\text{Lower number}} = \frac{3}{5} = \frac{3 \times 100\%}{5} = \frac{300\%}{5} = 60\%$

Examples

Example 1

The following table shows the units of penicillin produced by a fungus.

Sampling times (hours)	0	20	40	60	80	100	120
Concentration of penicillin (units per cm^3)	0	0	0.8	4.5	7	7.8	8

What percentage of the final concentration had been produced after 80 hours.

Step 1 – Identify the upper number in the fraction. At 80 hours = 7 units.

Step 2 – Identify the lower number in the fraction. At 120 hours = 8 units.

Step 3 – Write down the fraction = $7/8$.

Step 4 – Convert the fraction to a % by multiplying by 100% = $7/8 \times 100\%$.

Step 5 – Calculate value = **Answer** = 87.5%.

Example continued ➢

***Example** continued*

Example 2

Table 17.5 shows the energy used per hour by Justin Tyme for different activities.

Activity	Energy used (kJ/hour)
Sleeping	400
Sitting	500
Walking	1000
Swimming	1700
Running	2800

Which activity uses 150% more energy than sleeping?

In this example you have to **work in reverse** by converting a percentage to a fraction.

This means you **divide** by a 100% instead of multiplying.

Step 1 – Convert 150% to a fraction $= \frac{150}{100} = 1.5$.

Step 2 – Calculate by how much 1.5 is greater than the energy used in sleeping $= 400 \times 1.5 = 600$.

Step 3 – Calculate total for sleeping + 600 = 400 + 600 = 1000 kJ/hour.

Look at Table 17.4 for the activity that uses 1000 kJ/hour. **Answer** = Walking.

Example 3

More difficult questions ask you to calculate percentage increase and/or decrease.

A cylinder of potato tissue was weighed before and 1 hour after being placed into a salt solution.

Initial mass of potato cylinder = 10 g.

Final mass of potato cylinder = 9.4 g.

Calculate the percentage decrease in mass of the potato cylinder.

Remember in calculating a percentage **create a fraction**.

Step 1 – Identify the upper number in the fraction = Decrease = 10 – 9.4 = **0.6**.

Step 2 – Identify the lower number in the fraction = Mass at start = **10**.

Step 3 – Write down the fraction $= \frac{\text{Decrease}}{\text{Start}} = \frac{0.6}{10}$.

Step 4 – Convert the fraction to a % by multiplying by 100% $= \frac{0.6}{10} \times 100\%$.

Step 5 – Calculate value **Answer** = Decrease of 6%.

Ratios

A ratio compares two or more quantities/values in a **particular** order.

Ratios represent **relative** and not absolute quantities.

The word 'per' is used in ratios as is the phrase 'for every'.

For example, if the ratio of boys to girls in a school is 2 : 3, then this could be 20 boys **per/for every** 30 girls or 40 boys **per/for every** 60 girls, etc.

Method

Step 1 – Identify the value for the group that is **mentioned first**.

Step 2 – Identify the value for the group that is mentioned **second**.

Step 3 – Write down the values in the **correct order**.

Step 4 – Divide both values by the **smallest** value.

Example

Example 1

Samples of invertebrates were collected in two different areas.

Type of animal	Area	
	Bushes	Playing field
flies	60%	50%
beetles	35%	36%
spiders	5%	4%
ants	0%	10%

Calculate the ratio of flies to spiders in the bushes area. Show your answer as a **simple whole number**.

Step 1 – Value of 1st group = flies in bush area = 60.

Step 2 – Value of 2nd group = spiders in bush area = 5.

Step 3 – Values in the correct order = 60 : 5.

Step 4 – Divide both values by the smaller value $\frac{60}{5} : \frac{5}{5}$ = 12 : 1.

Answer = 12 flies : 1 spider.

Example 2

Examiners make questions more difficult by asking you to extract information from more complex data or by giving three different values or using numbers that do not divide to give whole numbers.

Example *continued* ➢

Example continued

Grass seeds were grown in separate beakers under two different conditions.

The lengths of the roots were measured every 5 days over a period of 25 days.

Day	Average length of roots (mm)	
	Beaker A	Beaker B
0	0	0
5	13	9
10	15	13
15	19	13
20	22	14
25	30	18

Calculate the simplest **whole number** ratio of average length on day 25 for the roots of the plants in Beaker A to those in Beaker B.

Step 1 – Value for 1st group = Average length of roots in Beaker A = 30.

Step 2 – Value for 2nd group = Average length of roots in Beaker B = 18.

Step 3 – Values in the correct order 30 : 18.

Step 4 – Divide by smaller value = $\frac{30}{18} : \frac{18}{18}$ **= 1.66666 : 1.**

This is NOT a whole number answer.

How do we get a whole number answer?

Restart at Step 4. Divide each side by numbers that give whole number answers.

If both numbers are even then divide by 2. This gives 15 : 9.

Can this be made smaller?

Both numbers can be divided by 3. This gives 5 : 3. **Answer** = 5 : 3.

Hints and Tips

Buy the Past Papers book and first identify and answer some questions relating to selecting information, then answer questions on presenting information and finally answer questions on processing information.

If your calculator gives an answer with three or more decimal places then most likely you have the wrong answer.

Chapter 18

PRACTICAL ABILITIES

The National Course Specifications for Intermediate 2 describes Practical Abilities as follows:

Questions relating to each of the following points will be included in the course examination in order to test the candidate's ability to:

4 *Plan and design experimental procedures to test given hypothesis or to illustrate particular effects. This could include identification of controls and measurements or observations required.*

5 *Evaluate experimental procedures in situations that are unfamiliar, by commenting on the purpose of the approach, the suitability and effectiveness of procedures, the control of variables, the limitations of equipment, possible sources of error and/or suggest improvements.*

6 *Draw valid conclusions and give explanations supported by evidence or justification. Conclusions should include reference to the overall pattern to readings or observations, trends in results or comment on the connection between the variables and controls.*

7 *Make predictions and generalisations based on available evidence.*

Terms often used in experimental work are **accuracy, validity** and **reliability**.

Accuracy is the **exactness of a measurement**. To measure time would you use a sundial or a stopwatch?

Validity is the '**fairness**'/'**correctness**' of a procedure. If you were investigating the effect of temperature on the rate of enzyme action by setting up a series of water baths at different temperatures then all other variables must be kept the same.

Reliability is the '**believability**' of results obtained. You would doubt the reliability of results if the reading of the measurements was inaccurate or if all other variables had not been kept the same or the experiment was carried out once only.

A pathway into practical abilities would be to take you through a typical Experimental Situation question together with a commentary.

Example

Figure 18.1 shows the apparatus used in an investigation to measure the rate of photosynthesis of the pondweed *Elodea* at different wavelengths of light.

Photosynthesis was measured by the volume of oxygen gas produced.

A water bath was used to keep the temperature constant throughout the investigation.

The method used in the investigation is outlined below.

A coloured filter was placed in front of the light source.

The *Elodea* was left for 5 minutes.

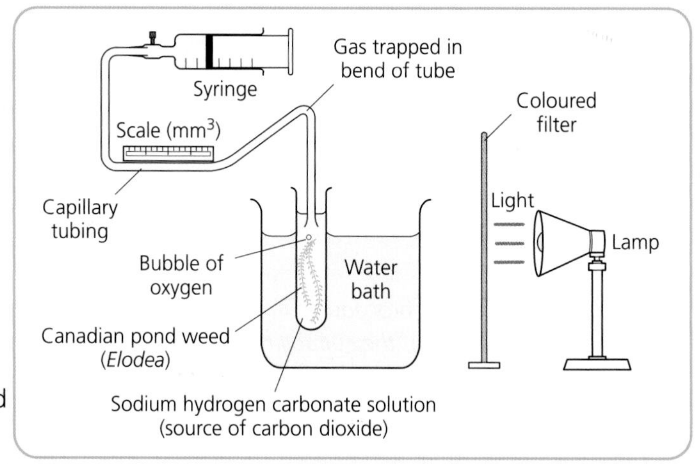

Figure 18.1

The syringe was pushed inwards to empty the capillary tube of gas bubbles.

The *Elodea* was left for 5 minutes.

The syringe was used to draw the trapped gas that was collected onto the scale.

This procedure was repeated three times.

This method was repeated for each coloured filter.

The table shows the results from the investigation.

Coloured filter used	Wavelength of light (nanometres)	Average volume of gas evolved (mm^3/15 minutes)
Violet	400	5.0
Indigo	450	5.8
Blue	500	7.4
Green	550	2.2
Yellow	575	3.4
Orange	600	5.0
Red	650	8.0

a) Describe the pattern in the change of the rate of photosynthesis over the range of wavelengths of light used in the investigation. *(2 marks)*

What does the question ask?

1 For a description,

2 of the pattern in the change of the rate of photosynthesis,

3 over the range of wavelengths used.

How to answer

1 Rate of photosynthesis is expressed as average volume of gas evolved.

2 The range of wavelengths from the table is 400–650 nanometres.

3 Read down and across both columns of the table. Use increases and/or decreases.

Answer

- As wavelength increases from 400 to 500 nanometres the rate increases.
- As wavelength increases from 500 to 550 the rate decreases.
- As wavelength increases from 550 to 650 the rate increases.

3 correct = 1 mark, 2/1 correct = 1 mark

b) The lamp gives out heat energy as well as light energy. Explain how the validity of the procedure would have been affected if the test tube containing *Elodea* had not been placed in a water bath. *(2 marks)*

How to answer

1 The water bath maintains a constant temperature. This is validity (correctness).

2 The lamp gives out heat energy and this increases the temperature. This removes the validity of the procedures.

3 The increase in temperature affects validity by increasing the rate of photosynthesis.

Answer

IT IS BECAUSE without the water bath the heat energy from the lamp would increase the temperature in the test tube – **1st mark AND** this would lead to an increase in the rate of photosynthesis – **2nd mark**.

c) Describe the control experiment that should be set up for this investigation.

How to answer

1 A control should be identical in every way to the experiment except for the factor that caused the change.

2 In biology experiments this factor is usually the living material.

Elodea must have caused the change.

Elodea must be omitted or replaced by non-living material.

Do not give a full description of the set up. Use the phrase '**Identical apparatus**'.

Answer = Identical apparatus without *Elodea* OR

Identical apparatus and *Elodea* replaced by plasticine (non-living material but taking up the same volume as the *Elodea*).

d) Explain why it was good technique to wait 5 minutes after the insertion of each coloured filter before collecting the gas.

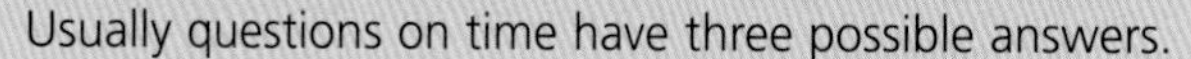

Hints and Tips

Usually questions on time have three possible answers.

1 Gives time for the change to go to completion.

2 Gives sufficient time to obtain a measurable value.

3 Allows time for the living organism/material to adjust to the new conditions.

Answer

Allows time for the *Elodea* to adjust to the new conditions.

e) How could the apparatus be modified in order to increase the accuracy of the measurement of the volume of gas given off at each colour.

Accuracy is related to the precision or exactness of the measurement. In this experiment the volume is measured in the tubing by reading the value from a scale.

Accuracy must be related to either:

1 The scale – divisions could make it difficult to obtain accurate readings. **OR**

2 The tubing used to collect the gas. If diameter is too wide a large volume is required for accurate measurable changes.

Use a scale with smaller divisions. **OR**

Use capillary tubing with a smaller diameter.

f) To improve the reliability of results the experiment was repeated five times at each wavelength of light. The results at 625 nanometres are shown in the table below.

Trial	1	2	3	4	5
Results (mm^3/15 minutes)	7.02	6.98	0.20	6.90	7.0

The result of trial 3 differs from the others. Suggest one possible source of error that could account for this result.

How to answer

The volume is lower than expected. And this is due to either less gas was given off or less gas was collected.

Answers include the following:

Wrong coloured filter used or lamp had been dimmed, or lamp was moved further back or some gas by passed the collecting funnel.

g) Identify three variables that must be kept the same at each wavelength of light in order to make the results obtained valid. *(2 marks)*

How to answer

Examine the set up to see which factors MUST be kept the same.

These would include:

1 Distance of lamp the same.
2 Light intensity of lamp the same.
3 Source of carbon dioxide the same.
4 Sample of *Elodea* the same.
5 Time for collection of gas the same.
6 Volume of water the same. **Any three from six.**

Example

Example 1

An investigation was carried out into the response of a plant shoot to directional light and gravity.

The experimental set up and results are shown in Figure 18.2.

What conclusion can be drawn from the results?

How to answer

From experiment 1, the shoot responds to the direction of light and towards the direction of gravitational pull.

From experiment 2, it grows away from the direction of gravitational pull in the absence of light.

Answer

The response to growth towards the direction of light is greater than the response to growth away from the direction of gravitational pull.

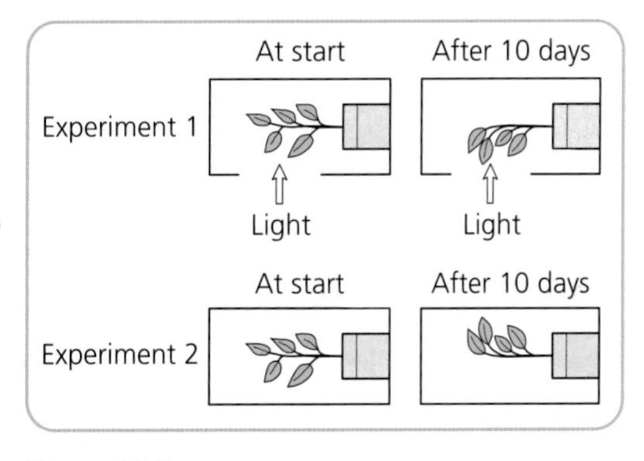

Figure 18.2

Example 2

Some students were asked to design an investigation to compare the rate of production of oxygen gas by two different species of water plant.

Figure 18.3 shows the proposed procedure.

Identify two changes to the procedure that would ensure that a valid conclusion would be made. *(2 marks)*

How to answer

Validity is about 'correctness' of procedures. How do the set ups differ/vary?

Example *continued* ➢

Example continued

Answer

1 Size/Mass/Surface area of leaves of plant should be the same.

2 Distance of lamp from plants should be the same. *(2 × 1 mark)*

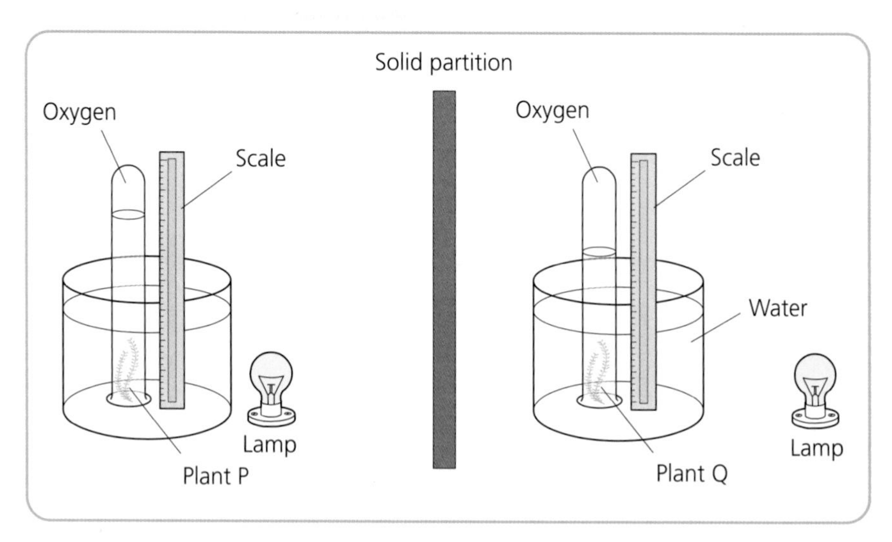

Figure 18.3

Example 3

Groups of students carried out an investigation into the effects of competition.

Four trays containing 15, 30, 45 and 60 seeds were set up.

The trays were watered regularly to allow germination to take place.

After several days the seedlings were observed and the number with healthy green leaves was noted.

***Example** continued* ➢

Example continued

The results are shown in the table.

Number of seeds in each tray	Number of seedlings with healthy green leaves	Percentage of seedlings with healthy green leaves
15	12	80
30	18	60
45	23	51
60	24	40

Predict the percentage of seedlings with healthy green leaves if 75 seeds were sown in an identical seed tray.

How to answer

In prediction questions you have to look for a pattern.

Number of seeds increases by 15 seeds each time.

Percentage decreases with each increase of 15 seeds.

From 80% to 60% = 20%. From 60% to 51% = 9%. From 51% to 40% = 11%.

There is no observable numerical value in the pattern other than it decreases each time 15 seeds are added.

The average decrease is 80% – 40% = 40% divided by 3 (number of additions of 15 seeds) = approximately 13%.

Answer

40% – 13% = 27%.

In the National Examination you would have been marked correct as long as you showed that the percentage was less than 40%.

Example 4

An investigation into the behaviour of blowfly larvae was carried out.

One blowfly larva was placed at X and then lamp A was switched on.

When the larva reached Y, lamp A was switched off and lamp B was switched on.

The path taken by the larva is shown in Figure 18.4.

(a) Describe the response of the larva to light.

You should notice the direction the larva moved in relation to the direction of light when lamp A was switched on and when lamp B was switched on and lamp A switched off.

Example continued ➢

Example continued

Answer

Larva move away from the light.

(b) Suggest one change to the apparatus that would confirm that the response was due to the direction of the light and not the heat from the lamp.

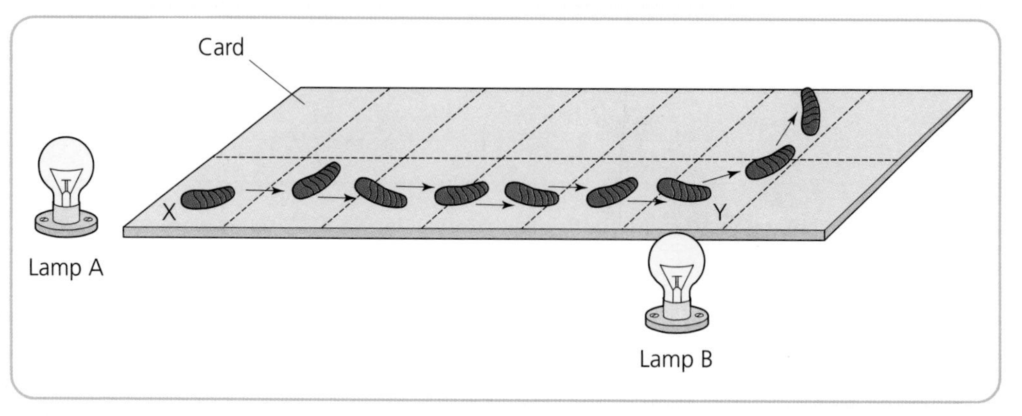

Figure 18.4

The clue is in 'not the heat from the lamp'.

Answer

Use lamps that do not give off heat OR place a glass slab between the lamp and the larva.

(c) Describe one way in which the reliability of the results would be improved.

Reliability is the '**believability**' of results obtained. If validity of procedure and accuracy of measurements are without question then the only answer is by replication.

Answer

Repeat the investigation several times OR use several larva rather than a single larva.

Sometimes the question is 'Why should you repeat the investigation several times?'

Answer

To improve the reliability of the results.

Hints and Tips

Buy the Past Papers book and identify and answer questions that relate to Practical Abilities.

Be clear of the differences between accuracy, validity and reliability.

When discussing the set up of the control remember that identical apparatus would be used but without the factors that caused the change.